The American Military Ethic

The American Military Ethic

A MEDITATION

James H. Toner

FOREWORD BY
Robert L. Wendzel

PRAEGER

New York
Westport, Connecticut
London

Library of Congress Cataloging-in-Publication Data

Toner, James Hugh, 1946–
The American military ethic : a meditation / James H. Toner ; foreword by Robert L. Wendzel.
p. cm.
Includes bibliographical references and index.
ISBN 0-275-94195-7 (alk. paper)
1. Civil-military relations—United States. 2. Military ethics—United States. 3. United States—Armed Forces—Officers. 4. Military education—United States. 5. Toner, James Hugh, 1946– I. Title.
UA23.T65 1992
355'.007'073—dc20 91-28747

British Library Cataloguing in Publication Data is available.

Library of Congress Catalog Card Number: 91-28747
ISBN: 0-275-94195-7

First published in 1992

Praeger Publishers, One Madison Avenue, New York, NY 10010
An imprint of Greenwood Publishing Group, Inc.

Printed in the United States of America

The paper used in this book complies with the Permanent Paper Standard issued by the National Information Standards Organization (Z39.48—1984).

10 9 8 7 6 5 4 3 2 1

For Rebecca
and Our Love Renewed
and to our sons
Chris, Pat, Jamey

Pax Domini Sit Semper Vobiscum

The nature of armies is determined by the nature of the civilization in which they exist.

—Sir Basil Liddell Hart

CONTENTS

Foreword ix
by Robert L. Wendzel

Preface xi

Introduction 1

1. Reveille: Basic Combat Training and Advanced Individual Training (Infantry) 9

2. Preceptory: Infantry OCS and Airborne School (Fort Benning, Georgia) 37

3. Second Lieutenant 65

4. First Lieutenant 97

5. Captain 123

6. Foundations of Professionalism: Liberal Education and ROTC 153

7. PME: Professional Military Education 175

8. A Bright, Shining Truth 201

9. Rhapsody in Blue: The American Military Ethic 231

Recommended Readings 257

Index 263

FOREWORD

Most authors claim their books are quite unusual, but most really are not. Jim Toner's *The American Military Ethic: A Meditation* actually is, in several ways.

To begin with, Toner very effectively combines autobiographical material (part one) with traditional scholarship (part two). This is difficult to pull off without the product appearing to be two only partly related pieces glued together. But he does this beautifully, making the transition from his firsthand reminiscences about his army training and service as a young officer to a mature scholarly analysis of key issues in military education and ethics.

Second, a significant portion of this book deals in an unusual way with a largely unexplored subject: the personal experiences of a junior officer in the U.S. Army in Europe during the Vietnam War. If as a nation we paid too little attention to the Vietnam War veteran, and we did, we paid even less to Vietnam-era veterans serving elsewhere, who *also* served their country well. Toner discusses the problems he observed in the peacetime army as a company-grade officer, the kinds of problems with which junior officers in peacetime must so frequently deal, but which receive little public attention except as media events. Written in a very entertaining style, many of Toner's accounts are amusing, but the young officer or cadet (and those who want to understand or help them) will find many of Toner's observations about leadership worth careful thought.

I have mentioned Toner's company-grade junior officer status a couple of times, because one of his express purposes in this book is to encourage a dialogue between junior and senior officers. When he discusses the military ethic and military education later, we see a more senior perspective. This, I think, has some real possibilities.

The education-ethics linkage just mentioned is a third outstanding feature. Toner believes strongly that in the American military, competence and ethics are symbiotic and that good leadership includes both knowledge and virtue. He also argues that competent ethical senior military leadership must be based on a firm undergraduate education, which subsequent professional military education then will extend and expand. He believes further in the "sacred" character of the American military ethic, and that a military force without ethical understanding is little more than an armed mob. This is strong stuff! Toner makes an impassioned plea for a considerable philosophical and ethical foundation in the education of American officers. Practicing what he preaches, at the Air War College he regularly teaches a course entitled "Command and Conscience."

Since his own time on active duty and his graduate school days, Professor Toner has been involved in scholarly analysis and the education of military officers, putting his ideas into practice. After he earned his Ph.D., his first teaching position, at Notre Dame, involved instructing a class on international relations, primarily for Navy ROTC midshipmen. He taught for nearly fifteen years at Norwich University in Vermont, one of four non-federal "essential military colleges" in the United States. He is currently serving as Visiting Professor of International Relations at the Air War College, one of the country's five senior professional military colleges.

The American Military Ethic: A Meditation is a well-written, well-researched, and at times provocative study of the creed that binds U.S. military professionals together. Some readers will agree very strongly with most of what is in this book; some readers will quarrel with more of what is in this book; but all readers will benefit from its entertaining but careful and thoughtful reflections on, and analyses of, some subjects of enduring and fundamental importance.

Robert L. Wendzel
Educational Advisor to the Commandant,
U.S. Air War College

PREFACE

> Far and away the best prize that life offers is the chance to work hard at work worth doing.
>
> —Theodore Roosevelt, Labor Day Address, Syracuse, New York, 1903

When the Korean War erupted in June 1950, I was four years old. My earliest memory of anything military was my fear in those days that my father, who had been a Navy officer during World War II, would be recalled to active duty—a fear often expressed during that three-year war by my mother. Although I did not dislike the military, I knew that I did not want my father to go away, and I knew that the military could recall him. After all, they had recalled Ted Williams, the famous Red Sox slugger, who had flown Marine jets in World War II (after hitting .406 in 1941). In the early 1950s, boys in western Massachusetts, where I grew up, were more aware of Ted Williams than of things military. On the other hand, I liked it when my father marched in the Memorial Day parade, even though he was not fond of marching. We worked out a deal whereby he would march every other year; it seemed fair.

My Dad was not recalled by the Navy. But in 1954, after his graduation from Monson (Massachusetts) High School, my cousin

Bill joined the Navy and served for four years. That was the extent of my association with the military. In the 1950s, however, it was generally understood that boys, eventually, would do a tour of duty in the military, probably after being drafted. Our model, at least in that respect, was Elvis Presley, who served his Army tour after getting his draft notice. Although I was not opposed to the military, I was in no hurry to join. In fact, one of my criteria in choosing a college was that it not have ROTC, for, in days of yesteryear, most U.S. colleges had ROTC programs that were, at least for a year or two of the undergraduate curriculum, compulsory for males.

As a 1950s Catholic, I grew up listening to Bishop Sheen preach against the evils of communism, and my parents were persuaded that Senator Joseph McCarthy of Wisconsin was, at least for a time, making a pretty solid case against the communist conspiracy within the councils of our government. I recall showing my father a color chart of the flags of the world and asking him which one he liked the best. I had been thinking of the various colors and patterns, and I felt somehow a little unpatriotic when my first choice, unlike his instantaneous selection, had not been the stars and stripes.

Growing up in the 1950s can hardly be explained today. You practically had to be there in order to understand it! In 1957, for example, I was shocked and chagrined at the news that the Soviet Union had launched *Sputnik.* I could not understand how the United States could be "defeated" in that regard. The television situation comedy *Happy Days* tried in its own inane way to convey some of the flavor of the 1950s. Of course, the 1950s were not at all happy days for thousands of Americans mired in poverty or deprived of their civil rights because of their color. Many of the problems perhaps endemic to the human species were as abundant then as now. But somehow the 1950s did, on balance, seem to be more carefree. We were not at war and we were universally regarded, for good reason, as a model of excellence around the globe. And if Ike and Mamie seemed a little stodgy in the White House, we had the glamor and glitter of stage and screen and sports to offset our political blandness.

I mention all this to underscore the fact that mine was not a military family. I was an only child in a lower-middle-class family growing up in a small town in western Massachusetts in what, in retrospect, appears to be a relatively halcyon time. It is almost certainly true that in the three decades that have passed since the end of the 1950s, America, American values, and American dilemmas have changed profoundly. My goal in this book is to provide one perspective on those changes. To do that, I must find a lookout of sorts from which to survey the scenery, and I have chosen to use as this lookout the American military, so often at the center of difficulty and debate these past thirty years. This is not an account that seeks to be "everyman's story." Nor in the pages that follow do I attempt to present myself as distinguished or heroic. I wasn't and am not. That every man is the hero of his own story may frequently be true, but I suspect that, in these accounts, the exception to that rule will prove valid.

The purpose here is not to provide a definitive scholarly inquiry into the nature of American society, for this is not a work of sociology. Nor is it to provide another was story about Vietnam, for I never served in Vietnam. In my undistinguished military service, I won exactly one medal that every soldier wins—it's called the fire watch medal, and it's given to soldiers who serve a certain number of days on active duty in recognition of their time on guard duty during that period of service—and one badge (the parachute badge).

I do have, however, a number of long-pondered stories to tell about the U.S. military. In my judgment, these stories fit into a kind of mosaic, offering a necessarily unique perspective about the imminent changes in the U.S. military and its parent society that, as 1959 segued into 1960 and the subsequent Age of Aquarius, hardly anyone could have prophesied. I say "necessarily unique" because my stories involve personalities and episodes whom and which only I know. What follows in these accounts are things that are true and that are presented here as factually and as faithfully as diaries and memory will permit. "Only the names have been changed to protect the innocent." In some cases, when no point or purpose would have been served in using real names, I have used made-up names. But all the accounts are derived from actual occurrences, however fantastic some of them may seem.

There is one over-arching purpose to this volume and, at the outset, I must state that I know my accounts cannot fully satisfy that purpose. The Vietnam War continues to be fought in scholarly forums[1]; in popular books, movies, and TV shows; in bars and at ballgames. Like the Civil War, it will not be over until all its veterans are deceased—if even then. What its lessons are for national security is puzzling enough; what its effects are in the lives of its veterans perhaps only God knows. To be sure, the Vietnam veteran was treated miserably by his countrymen when he returned from his tour in Vietnam.[2] Some amends, not enough, have been made. Some of the protestors, too, have had their day. Those who served in conscience—and those who refused service in conscience—were, let it be said, patriots alike. In their own way, they were brothers in the service of ideals.

Of course, there were Vietnam veterans who did not serve their country well or loyally. And there were draft dodgers who were cowards. And one can argue about whether some of the protestors served high purpose or, in some of their acts, committed treason. But those arguments are not in these pages. Instead, these pages seek to tell only part of the story of another group, whose trials have not been described. This is, in small part, the story of the *Vietnam-era* veteran. Many thousands of Americans proved to be true and faithful servants of the Republic without ever spending a day in Vietnam. They may have served in CONUS (within the United States), in Europe, in Korea, or elsewhere; technically, they are not Vietnam veterans. But neither were they draft dodgers.

A little while ago, I talked with a man I know in Vermont, where I now make my home. Our conversation somehow turned to things military, and he told me, almost by way of confession, that he had been in the Vermont Army National Guard during the Vietnam War. More than a hint of shame emerged as he told me that he had made a mistake. "I should have been in the Regular Army," he said, "instead of in the Guard." Those who protested the war on grounds of honest conscience and genuine patriotism, within the boundaries of legal and ethical propriety, have their comfort, and they should. Those who served in Vietnam on grounds of honest conscience and genuine patriotism, within the boundaries of legal and ethical propriety, have their comfort, too—as well they should. Both groups have their own demons to contend with. But the Vietnam-era veteran, in which category my Vermont friend is, belongs to neither world. These pages are not "his" story. But in these pages is one such story, and it may shed some light, not just on that species of U.S. veteran, but on a "peacetime" Army in wartime.

In 1959, I was thirteen and, on 31 December that year, aware that dramatic changes would occur in the 1960s. How dramatic I hardly knew! In 1959, I was a fairly typical teenager. In the next thirty years, the world was to turn over many times, taking me and my generation with it. All the institutions I knew that winter night in 1959 were to change, some of them violently. In the 1960s, boys began to challenge the draft and Mohammed Ali—not Elvis—became the model for many. The United States, instead of being the paragon of virtue, became for many the incarnation of evil. Stars of stage and screen became politicized: witness John Wayne on the one hand and Jane Fonda on the other. Sports stars were not flying Marine Corps jets; many were fighting (and some losing) their battles with alcohol and drugs. Some of the movie stars and athletes had salaries so immense that a kind of general corruption seemed always at the edge of Hollywood and of sports. The Stars and Stripes was burned and spit upon—by Americans. ROTC was summarily expelled from many campuses. A bland American politics became explosive in the 1960s as many young leaders of the 1950s reached the heights—and met their fates—in the 1960s.

In 1956 I remember watching the Democratic national convention on TV. My mother was thrilled when, for a time, it looked as though a young Irish Catholic from Massachusetts might become the running mate of Governor Stevenson. Senator John F. Kennedy did not receive that nomination. But four years later, of course, he did land the presidential nomination of his party. Three years after that, he was dead, signaling the real beginning of the 1960s. In 1955 a young black leader in Alabama began his rise to fame as the very personification of the civil rights movement in the United States. Thirteen years later, he was tragically gunned down in Memphis. Also thirteen years later, I was about to graduate from college, preparing to listen to our commencement speaker, Vice President Hubert Humphrey. Humphrey did not come to campus that day. He turned down the engagement after learning that his rival for the Democratic presidential nomination, Bobby Kennedy, had also been tragically cut down.

Life can be ironic. For one who never sought military connections—and for one who thought the military was well past him after his honorable discharge was granted—I have spent most of my professional career as teacher and scholar in dealing with the military. I teach at Norwich University, the nation's oldest private military college. As I write these words, I am engaged for two years teaching at the Air War College, the highest professional educational institution of the U.S. Air Force. In a very real sense, my principal professional preoccupation has been the education of officers. That will be the focus of chapters on ROTC and PME (professional military education). The irony is that, as an eighteen-year-old, I deliberately chose to attend a college that did not offer ROTC.

I believe that there is something sacred about the American military ethic (AME) and the way we in this country go about civil-military relations. Those themes, which are connected, are the subjects of the book's final two chapters. The autobiographical chapters (1–5) are, of course, first-person accounts. The chapters on military education and on the military ethic (chapters 6–9) are "academic" in the sense that they are documented fairly thoroughly. The notes to all the chapters are intended to encourage further thinking and reading about the points and issues raised. The book concludes with a "Recommended Readings" section.

There are a number of people whom I must thank for their help and cooperation as I wrote this book. Primary among them is my wife Rebecca, without whose encouragement I could never have undertaken this project. My sons Chris and Pat, both born in Army hospitals, have their own Army stories now, and I am proud of the fact that they are willing to serve their country. My son Jamey, Vermonter that he is, has gamely adjusted to his father's "tour of duty" in central Alabama. He worries, I think, that he is coming to like it. For years I have inflicted on my family the querulous nature of someone at odds with the times, trying, I hope manfully, to understand that he might teach more cogently; write more coherently; speak more clearly; perhaps, one day, live more compassionately. My successes are yours; my failures, my own. I am reminded of the epitaph on a tombstone in Arizona. "Here lies Bill Jones. He done his damnedest." Maybe I can, with your patience and understanding, try harder not to try too hard.

My thanks go, as well, to Norwich University, which has provided a secure haven in which to teach and write—and even, occasionally, to coach—these nearly fifteen years. The motto of Norwich ("I will try") seems quaint to many. I take it, however, as an imperative and have tried, with varying degrees of success, to "resemble that remark" to my students. Many colleagues, many students, many friends have helped to shape the thoughts in these pages. My love of the Norwich ideal is diminished only to the extent that the institution occasionally neglects its unique experience and educational mission. The French historian Fustel de Coulanges (1830–1889) took for his motto, "Quaero"—"I seek to learn." That shibboleth encapsulates what any genuine institution of higher learning ought to be about. Officers of the future cannot know what to do unless they first know what to be, knowledge in part derived from serious study of serious academic subjects at serious colleges. If and when such a college misapprehends its true purposes, it effectively betrays both its past and its future.

The administration, faculty, and students of the U.S. Air War College (Maxwell Air Force Base, Montgomery, Alabama) have been cordial and challenging and encouraging. For his counsel and kindness, I am especially grateful to Dr. Robert Wendzel, Educational Adviser to the Commandant of the Air War College. My time at the Air War College has been intellectually stimulating and greatly enjoyable. The people at the Air War College, men and women of character and competence and commitment, are what the American military ethic is all about. I am honored to have been in their company. (Indeed, as I wrote some of these words their brother officers, many of them graduates of the Air War College, were in harm's way over the skies of Iraq and Kuwait.) Without the wonderful opportunities for research and teaching extended to me by the Air War College (and without the excellent facilities of the Air University Library), I could not have accomplished this project. Of course, the opinions and judgments in this book are my own and are not necessarily shared by the Air Force, Army, or U.S. Government.

I gladly acknowledge and greatly appreciate the advice and encouragement of Dan Eades, my editor at Praeger. Although I did not always accept his editorial admonitions and adjurations, I frequently did—and the book is better because of his interest and experience. Finally, I thank the readers of this book. You will not find in its pages very many answers to the intractable problems of life. You may encounter, I hope, some worthy questions. Mr. Justice Louis Brandeis (1856–1941) used to be asked by his mother every day when he came home from school, "Louis, did you ask any good questions today?" If you think of good questions, prompted by reading this book, my efforts are well served by having written it. The introduction will suggest to you my questions and purposes as I set out on this project.

NOTES

1. See, for example, Bob Buzzanco, "The American Military's Rationale Against the Vietnam War," *Political Science Quarterly* 101 (1986): 559–576.

2. See David Donovan, *Once a Warrior King: Memories of an Officer in Vietnam* (New York: McGraw-Hill, 1985).

The American Military Ethic

INTRODUCTION

> It is as fatal in politics to ignore power as it is to ignore morality.[1]
>
> —E. H. Carr

> The day may dawn when fair play, love for one's fellow men, respect for justice and freedom, will enable tormented generations to march forth serene and triumphant from the hideous epoch in which we have to dwell. Meanwhile, never flinch, never weary, never despair.[2]
>
> —Winston Churchill (1955)

This is a book about leadership written by an author who has never led more than a college course, or a high school baseball team, or an Army company (and that not in combat). It is a book about ROTC by an author who wanted nothing to do with ROTC when he was an undergraduate and who deliberately enrolled in a college that did not have ROTC. This is a book about the education of lieutenant colonels and colonels, many of whom will be future generals, written by an author who barely made the rank of Army captain. This is a book about the American military ethic written by an author who, with the exception of one tour of active duty (1968–1972), has always been a civilian.

This is not so much a book about answers to problems in leadership, combat style, military education, and the military ethic as it is an assemblage of questions about those subjects and ideals. What is the purpose of basic training in the Army and what occurred in basic during the late 1960s? Thousands of Americans have gone through Officer Candidate School, but there is very little published about it. Was it effective? The infamous Lieutenant Calley is an alumnus of Infantry OCS. Was he typical? What was it like to go through airborne school? What was the United States Army, Europe (USAREUR) like in the early 1970s? What was it like being in charge of nuclear weapons convoys? How effective was Army discipline in the late 1960s and early 1970s? What lessons, if any, can a former non-career company-grade officer pass along to younger officers today? Does ROTC "work"? How can it be improved? What is PME and what are the war colleges? Is there danger of a military coup in this country? Has the military been "civilianized"? What is meant by the term "American military ethic"? Is it important? Why?

This is a book about leadership by an author who has studied and thought and written about leadership all his life—and whose main conclusion is that of President Franklin D. Roosevelt: "A good leader can't get too far ahead of his followers" (1940). This is a book about ROTC written by an author who for nearly fifteen years has taught at the college that pioneered the very idea of ROTC, Norwich University in Vermont, which is a military college. This is a book about professional military education by an author invited to be Visiting Professor of Political Science for two years at the U.S. Air War College (Maxwell AFB), which is one of five senior PME colleges in the United States. This is a book about the American military ethic by an author privileged to be, both as soldier and as scholar, in the company of men and women who *serve* that ethic and who, moreover, *are* that ethic.

Because this book attempts to combine "autobiographical" as well as "academic" portions, it is an unusual effort. In the autobiographical part, I relate a number of incidents and episodes I was personally involved in; the exalted American military ethic can mean very little unless it seeps down to the junior officer in the field. I was very junior, and I was in the field—at an isolated post in a "peacetime" Army. My job was to secure and to help maintain nuclear weapons. My story of mission and men is nothing compared with the life-and-death episodes related so well by other platoon leaders and company commanders.[3] But few officers, in fact, end up being combat leaders; most deal with other kinds of battles. Mine were with racism and drugs and illiteracy; these enemies, however, can be as mortal as any other national enemy of the United States. My goal is not to provide material for a guts-and-glory TV show—but rather to examine garrison leadership and the military ethic at a very tough time in a very tough place. I try, too, to be frank about my own errors in judgment and lapses in maturity. In the book's other parts, I try to examine education and ethics more "clinically" and academically. My goal is to relate all the chapters to the idea that the United States has an intellectual and ethical treasure in the military ethic, about which there is a true spiritual quality. For its part, the AME (American military ethic) depends upon PME (professional military education). If that sounds entirely too much like a military acronym, what I mean is this: for the American military ethic to flourish, we must have educated, ethical, morally resolute leaders. I champion, therefore, substantial ROTC/OCS education and rigorous professional military education, both of which must be rooted in the liberal arts and in character development. Not for nothing did the late Morris Janowitz observe that, "In short, every effective soldier, by his very task, must be a teacher."[4]

A number of critics would argue that as teachers and as moral exemplars, the American military failed bitterly during the period about which I reminisce and the period after. More than fifteen years ago, I argued somewhat along those lines myself.[5] But it was Rick Gabriel and Paul Savage who seem to have started the ball rolling in their controversial book *Crisis in Command.*[6] That volume is as trenchant as it is truculent. It said things that weren't very nice; it also said them in a tendentious manner. Both the substance of their research and their polemical manner were later repudiated by a number of disinterested scholars, as well as by deeply offended military officers.[7] Having known these two writers virtually since the appearance of that book (and as a graduate of Saint Anselm College, where they teach), I have defended them, *up to a point*, both in print and in private. Much of what they write is true; there is no circumventing that. And to the extent that they started an intellectual conflagration among those interested in military affairs, they deserve credit, not censure. Neither of these gentlemen, however, is diffident or fawning, and the withering fire they took (much of *that* deserved!) passed largely overhead. After all is said and done, Gabriel and Savage helped the military to examine its conscience; that is the first step on the road to a statement of contrition and to a firm purpose of amendment. Professors at a Catholic college might well be satisfied with that.

There are, and were, episodes of chicanery and foolishness along the road to reform. Gabriel, for example, rushed into the fray to defend Cincinnatus,[8] who blasted the Army officer corps for its performance in Vietnam. Posing as a combat veteran, "Cincinnatus" later turned out to be a chaplain who never served in southeast Asia. That might well have been all right, but he had misrepresented himself and refused even to reveal his own name. The literature of the Reform Movement is nearly as vast as that of the Vietnam War. As Gabriel's mistaken defense of Cincinnatus suggests, even the reviews of the literature are very uneven and sometimes unreliable.[9] More recent reform literature, however, is of better quality—that is, "less caustic and more cause." Books such as Arthur T. Hadley, *The Straw Giant*; Edward Luttwak, *The Pentagon and the Art of War*; James Coates and Michael Killian, *Heavy Losses*; Richard Halloran, *To Arm a Nation*; Gary Hart and William Lind, *America Can Win*; Bob Woodward, *The Commanders*—all seem more balanced and fairer than some of the earlier attacks on the military.[10] The literature of Vietnam, fictional and factual, is so large that no attempt will be made here to provide an overview.[11] But students of the military simply have to learn as much as they can about the Vietnam background to understand the military today.[12] Efforts at reform[13] seem to have culminated successfully in the Gulf War of 1991.

This book is not about the "Reform Movement" as such. Nor is it about Vietnam. It concerns the ethic, the philosophy, that is part—or ought to be part—of the mental "baggage" of every soldier, whether of long-term or short-term service. Books by Huntington,[14] Janowitz,[15] and Weigley[16] are the essential references for any volume of this sort. Indeed, they are indispensable to students of the subjects discussed here. Equally invaluable to a study of the military is the work of the late Hans Morgenthau, who wrote that "the struggle for power is universal in time and space and is an undeniable fact of experience" and, consequently, "[i]f the desire for power cannot be abolished everywhere in the world, those who might be cured would simply fall victims to the power of others." As the epigraphs to this introduction suggest, we require, for our national safety, appreciation of power as well as appreciation of ethics—a concept Morgenthau taught so well.[17] Morgenthau also explicitly taught the intellectual *sine qua non* of the American military ethic: "The armed forces are the instrument of foreign policy, not its master."[18] I gladly acknowledge my indebtedness to all of these scholars.

The American Military Ethic: A Meditation is divided into nine chapters and a "Recommended Readings" section. Chapters 1 and 2 discuss my training in the Army; chapters 3, 4, and 5 discuss the practical leadership problems I encountered after graduating from OCS. Chapters 6 and 7 deal with the education of officers, junior and senior. Chapters 8 and 9 examine the AME today.

More specifically, chapter 1 is an examination of basic combat training, through which have passed thousands of American men (and women) on the way to being "G.I.'s." Basic is, however, rarely examined in a serious light, in which the examiner asks what its purposes and expectations are. This chapter also provides a virtually unique view of Infantry Advanced Individual Training (AIT), through which our Army combat infantrymen must pass.

Chapter 2 explores the training of an Army infantry officer in the grueling six-month Infantry Officer Candidate School at Fort Benning, Georgia. OCS was intended both to turn out tough, competent platoon leaders and, of course, officers imbued with the military ethic. It's tragic that the best-known alumnus of Infantry OCS is Lieutenant Calley, for OCS officers, on balance, did a creditable job. A number of observers may not like the highly decorated Colonel David Hackworth, but no one (to my knowledge) has ever challenged his competence in the field or his personal bravery. "We concluded," Hackworth says, "that OCS officers statistically made better platoon leaders than ROTC or West Point graduates because they had more hands-on training."[19] This chapter will discuss why that may be true. It also explores the training of the U.S. Army Airborne School at Fort Benning. Instructors there are wont to suggest that there are those who have proved their worth in airborne school—and then there is the rest of humanity. That is probably a little bit of hyperbole.

Chapters 3, 4, and 5 examine my time as a lieutenant and captain in USAREUR, the U.S. Army, Europe. They deal with practical leadership problems encountered by the junior officers of that day in that army. These chapters are arranged topically rather than chronologically, and they describe problems with mission as well as with men. The American military ethic is not something for War College students (usually colonels and rising lieutenant colonels) alone. Indeed, troops are closest, by definition, to lieutenants. These chapters examine in a very practical way what I was taught in basic, in AIT, in OCS, and in airborne school that I used (or ignored) in the field with real troops (as opposed to practicing leadership in the if-it's-Monday-then-it's-your-turn-to-be-platoon-leader kind of instruction that cadets often receive).

Chapters 6 and 7 examine education, which constitutes the heart and the head of the AME. An Army second lieutenant will spend about a year and a half in the classroom during his first five years of service. By the end of a twenty-year career, an officer has usually devoted five or six years to study.[20] Today's military knows that without serious and substantial education, the junior and senior officers of our armed forces will be lost. They will not know what to do, for they will not know who they are. Any officer who cannot distinguish right from wrong and honor from shame is a ticking time bomb. A solid liberal arts education, either as a major itself or as a substantial complement to an education in, say, the sciences or engineering, should help the student to develop the ethical compass he will need to chart a wise course throughout his career and life.

Chapters 8 and 9 provide a look at the military ethic. Chapter 8 examines military justice and the idea of civil-military relations in this country, which is our "bright, shining truth" (a play on Neil Sheehan's biography of John Paul Vann). And chapter 9 looks at the two major jobs of the soldier—to kill and to die—exploring why ethics is so literally vital in the education of the officer and in the subsequent execution of his officerly duties. The title of the final chapter is "Rhapsody in Blue," for all the U.S. armed services have at least one blue uniform. Blue is the color associated with loyalty and piety, both of which are necessary in the world of the officer and gentleman. In the end, the American military functions to the extent that our society functions, for our military is a specialized community within our larger society. This last chapter examines models of the soldier, one suggesting that soldiers ought to be divorced from politics and the other suggesting (I think correctly) that officers must be highly educated in politics *and* diplomacy, as well as in the operational arts and sciences. Few soldiers will ever rise to being influential strategists, but those who do reach that level must comprehend more than ordnance and logistics.

The hope is that this book's combination of the personal and professional approach to the military can appeal to two audiences: the audience (such as students and cadets) that wants an experiential, first-person, base; and the audience (such as seasoned scholars and soldiers) that wants a more formal approach.[21] My goal is to encourage the two groups to talk with each other. It's not a bad thing for senior officers to remember their days as cadets or candidates or lieutenants. And it's not a bad thing, either, for aspiring cadets to think ahead, at least occasionally, to being senior leaders. It is indeed a very positive thing for academics to remember that soldiers often live in a world of mud and sweat—and sometimes blood; academic analysis can and should be closely related to reality. Finally, it is a very positive thing for practical soldiers to remember that theirs is, at its best, a very idealistic profession—and that if they are to be professionals, they need always to stay closely in touch with what it is they ostensibly profess, which is the American military ethic.

Finally, this book is not just for professional soldiers of either the long-term or short-term variety. It is intended, too, for the veteran who served with true faith and allegiance to the Republic, whether in Vietnam or elsewhere during the period the book is concerned with. S. L. A. Marshall had it right: "The military system of the United States is not committed to grinding out warriors *per se*, but to the training of men in such manner that they will be able to play a better part anywhere, and will find greater satisfactions in what they do."[22] In the end, whether volunteer or conscript, it is not *their* (the soldiers') armed forces. The uniforms worn by those soldiers, sailors, airmen, Marines, and Coast Guardsmen are ours. They represent us. Their philosophy is America's. They serve and they preserve our noble heritage, the soldier's faith, here called the American Military Ethic.

NOTES

1. E. H. Carr, *The Twenty Years' Crisis* (New York: Harper and Row, 1964), p. 97.
2. From a Churchill address of 1 March 1955 in *Parliamentary Debates*, 5th Series, Vol. 537, col. 1905.
3. See, for example, about World War II: Charles B. MacDonald, *Company Commander* (Washington, DC: Infantry Journal Press, 1947); and about Vietnam: James R. McDonough, *Platoon Leader* (Novato, CA: Presidio Press, 1985).
4. Morris Janowitz, *The Military in the Political Development of New Nations* (Chicago, IL: University of Chicago Press, 1964), p. 81.
5. James H. Toner, "The Military Ethic: On the Virtue of an Anachronism," *Military Review* 54 (December 1974): 9–18.
6. Richard Gabriel and Paul Savage, *Crisis in Command* (New York: Hill & Wang, 1978). See also Jeffrey Record, "Is Our Military Incompetent?" *Newsweek*, 22 December 1980: 9.
7. See Forrest Waller, "Are Officers Incompetents?" *Air University Review* 36 (Sept/Oct 1985): 73–82 and his notes.
8. See Gabriel's review of *Self-Destruction* (New York: Norton, 1981) in *Air University Review* 32 (July/Aug 1981): 96–97.
9. For a good review of seven books about reform of the military in and after Vietnam, see John P. Lovell, "No Tunes of Glory," *Indiana Law Journal* 49 (Summer 1974): 698–717.

10. Two earlier books, William Corson, *The Consequences of Failure* (New York: Norton, 1973) and Douglas Kinnard, *The War Managers* (Hanover, NH: University Press of New England, 1977), may be of interest.

11. Aside from the references provided in the notes to any serious Vietnam book, those interested in that war should consult Ronald H. Spector, *Researching the Vietnam Experience*, Analysis Branch, U.S. Army Center of Military History (Washington, DC: Government Printing Office, 1984). (I thank my Air War College colleague, Dr. Alexander S. Cochran, for calling this to my attention) and *Vietnamese Conflict: Lessons Learned*, Special Bibliography Series, Air University Library (Maxwell AFB, AL; December 1990).

12. See Drew Middleton, "Vietnam and the Military Mind," *The New York Times Magazine*, 10 Jan 1982: 34–37 and following.

13. See "The Fight to Change How America Fights," *U.S. News & World Report*, 6 May 1992: 30–31.

14. Samuel Huntington, *The Soldier and the State* (Cambridge, MA: Belknap Press, 1957).

15. Morris Janowitz, *The Professional Soldier* (Glencoe, IL: Free Press, 1960).

16. Russell F. Weigley, *The American Way of War* (New York: Macmillan, 1973).

17. Hans Morgenthau and Kenneth W. Thompson, *Politics Among Nations: The Struggle for Power and Peace*, 5th ed. (New York: Knopf, 1985), p. 38. Morgenthau says at one point that there are limitations on foreign policy that are rooted in transcendent moral principle. For example, a policy of mass extermination, however advisable from a strictly utilitarian position, is unthinkable—even though "expediency would counsel such a thorough and effective operation. The limitation derives from an absolute moral principle, which must be obeyed regardless of considerations of national advantage. A foreign policy of this kind, therefore, actually sacrifices the national interest where its consistent pursuit is the violation of a moral principle, such as the prohibition of mass killings in times of peace" (p. 252).

18. Ibid., p. 590.

19. David H. Hackworth and Julie Sherman, *About Face* (New York: Simon & Schuster, 1989), p. 593.

20. David Gergen, "Victory Can Teach Lessons," *Montgomery Advertiser*, 6 May 1991: 7A.

21. Two recent good texts are those of Loch Johnson, *America as a World Power* (New York: McGraw-Hill, 1991) and Sam Sarkesian, *U.S. National Security: Policymakers, Process, and Politics* (Boulder, CO: Lynne Rienner, 1989).

22. S. L. A. Marshall, *The Officer as a Leader* (Harrisburg, PA: Stackpole, 1966), p. 40.

1 REVEILLE: BASIC COMBAT TRAINING AND ADVANCED INDIVIDUAL TRAINING (INFANTRY)

> [I]f she [your country] leads you to war, to be wounded or die, this you must do, and it is right; you must not give way or retreat or leave your post, but in war and in court and everywhere you must do whatever city and country demands, or else convince her where the right lies.
>
> —Words attributed to Socrates by Plato in the *Crito*

> In no other profession are the penalties for employing untrained personnel so appalling and so irrevocable as in the military.
>
> —Douglas MacArthur, *Annual Report of the Chief of Staff*, 1933

In college, during my senior year, I became very interested in the Korean War. But my interest in that war—and in the war in Vietnam that was raging throughout (and subsequent to) my college career—was rather clinical. "To make nails do not waste good iron," a Chinese proverb has it; "to make soldiers do not waste good men." As I studied the war in Korea and debated the war in Vietnam, from 1964 to 1968, I did not suppose that I would be wasted. I thought that the Army was for the other guy. The point of this chapter is to detail the evolution of my thinking about that judgment and to set out the reasons that I joined the Army in 1968. In addition, the chapter describes the training in BCT (basic combat training)—through which millions of Americans have passed—eight weeks (in 1968) of sometimes rugged, sometimes silly, military training. After basic, soldiers go on to Advanced Individual Training in their MOS (military occupational specialty, or job). My AIT happened to be in infantry—which was the lot of thousands of U.S. "grunts" or ground troops who have fought our country's wars.

I discovered that the macrocosm of things I had learned about the Korean War had now funneled down to a very small microcosm indeed—me. In the end, the American military ethic depends upon the individual soldier who learns that ethic and who lives it, to one extent or another. For me the world of the scholar ended, not with a whimper, but with a bang. The banging sound I heard was reveille; I was in the Army. And my interest in the military was no longer just clinical; it was immediate, and personal, and not very antiseptic at all. I have learned to be very suspicious of Chinese proverbs.

I

On 14 May 1968, my twenty-second birthday, I enlisted in the United States Army. That was not something that, only a few months before that date, I had expected to do. I suppose my decision to join the Army had begun four years before the explosive year of 1968.

I wanted to go to college, but it looked as though my family would not have the money to send me, even though I was an only child. I had chosen the college I wanted to attend: Saint Anselm's College in Manchester, New Hampshire. It had met all my criteria: small, Catholic, liberal arts, in New England, with a good history department, with teacher training, no ROTC (I wanted no part of that!), no math requirement (I hated math), and no women (I did *not* hate women, but I did not want to attend a co-educational college). Saint Anselm's then had a nursing department, but it was not co-ed; in those days, not too many Catholic colleges were co-ed. I matriculated at Saint Anselm's in September 1964. By June 1968, when I graduated, the world was changing dramatically.

The world, for me, did not begin to change in the fall of 1964, for, although at eighteen I was not quite old enough to vote,[1] I strongly supported President Johnson over the challenger, Senator Barry Goldwater of Arizona. As an Irish Catholic, I thought of myself as an "inherent Democrat"; Johnson was to be the last presidential Democrat I supported. By 1972 and McGovern, I was puzzled how anyone could support the Democrats of "acid, abortion, and amnesty." How the world changes!

It was 1965 before I began to notice the changes or, perhaps more correctly, began to notice that I was not keeping pace with the changes. An earnest young man named Jim Salmon was, with me, a sophomore at Saint A's (as it was called). Salmon had been in a seminary studying for the priesthood; when he left, he decided to enroll at St. A's. He brought with him a great concern about the war and a deep conviction that he must burn his draft card—that he could not participate in the war in any way, even to the point of bearing a military registration card.

At the time, there was a draft or conscription. But students could gain the privileged 2-S status by attending college. The college's registrar would then notify the student's draft board of his attendance, and the draft board would approve the deferment. Salmon was deferred; but it was not enough for him. As I recall, Salmon was the second person in the nation to burn his draft card that fall. I believe that David Miller beat him to the punch.[2] But a number of people in southern New Hampshire supported the inchoate anti-draft, anti-war movement. At Berkeley in California, the Mario Savio free speech movement was under way. The time of the radicals was coming; the Age of Aquarius had begun.

I attended a debate one fall evening in 1965 at the theater at Saint Anselm's. A group of anti-war students, well before such had become chic, was calling upon others to join them in protesting the Vietnam War. Most students, still in short hair and conservative dress (we were then required to wear a coat and tie to class!), either actively disputed the nascent radicals or remained apathetic and ignorant. I recall a brief speech that night by Dr. Carel Jan Smit, an economics professor and naturalized American. In words I found to be strangely moving, Dr. Smit reminded the radical students that those who would not stand up for their country—such as the Oxford University students of the 1930s who had pledged never to fight for king or country—often had to change their minds when tyranny took advantage of their quixotic ideals. A truly decent and humane man, Dr. Smit even then might have been committing the error, prevalent in conservative circles, of confusing opposition to a given political strategy with lack of patriotism. In any case, this was the first taste I personally had of the national movement that was to sweep the country.

By the summer of 1966, the idea was in vogue that if you went to San Francisco, you wore flowers in your hair. I was not born on the Fourth of July. I knew politicians could blunder. And I recalled Carl Schurz's dictum: "Our country, right or wrong! When right, to be kept right;

when wrong, to be put right." This comment in the Senate, made in 1872, paralleled Plato's wisdom—one of the epigraphs to this chapter—from centuries before. (I repeat Schurz's wisdom in chapter 9, regarding it as entirely consistent with the American Military Ethic.) I was not a hopelessly naive young man. I too was questioning the wisdom of our involvement in Vietnam, albeit not on ethical grounds (about which I had little doubt). I simply wondered whether we had compelling geopolitical interests there. Then I went to the Union of Soviet Socialist Republics.

I was fortunate enough to be selected by an organization (the Citizens' Exchange Corps) in spring 1966 to spend a month (July 1966) in the USSR. It was a financial hardship for me; it meant taking a month off from work. But I could not let the opportunity pass, and they provided me with a half-scholarship to make the trip possible. In terms of what I guess I was supposed to take away from the trip, either it or I failed badly. My conversations with Soviet professors and students, with average Soviet adults and kids, led me to the conclusion that the Soviet regime was savage and dangerous. For example, during one class at Moscow State University, the professor, a Soviet historian of American history, inveighed on and on about American genocide in Vietnam, our treatment of Indians and Negroes, our widespread poverty, our lack of genuine political freedom, our incompetent capitalist-imperialist system, and the like. As a student, I was accustomed to asking questions, and I responded by asking him about Katyn Forest, about Boris Pasternak, about Stalinism, about Hungary in 1956, and the like. The question had been argumentative, it is true; but it had been stated with some aplomb (if I may say so). I will never forget his response. It was "Shut up and get out!"

I wish I could say I snapped off a brilliant rejoinder. In fact, I was terrified. I did not get out. I sat in my chair too scared to move. I offered no more debate. He ignored me. Later in the trip, I talked with a young student at Moscow State University who peremptorily informed me that the Berlin Wall had been built by West Berliners as a security fortress. Of course, I knew that the East Berliners, under Soviet orders, had built it five years before, not to keep West Berliners out, but to keep East Berliners in. When I protested along these lines, my antagonist challenged me to prove it. *Pravda* and *Izvestia* proved to be of little help! When that infamous wall was finally torn down twenty-five years after my contretemps, I thought of my rival: Did he, finally, know who had built that damned wall—and for what purpose? I hope so.

One evening in a cafe, I was sitting with a young couple (both civil engineers). After a long conversation, the woman went to the ladies' room. Her husband told me, *sotto voce*, that "My dream is one day to go to America. But do not tell my wife." We could talk about virtually anything—except politics, which they did not want to discuss. They thought things were "much better now" in the Soviet Union; their parents had told them about Stalin (who had died in 1953). The lack of freedom, the economic deprivation, the absence of serious discussions among serious people, the inability of the people to have access to news or to reliable historical accounts—all these things persuaded me that communism was not merely an intellectual "ism" but rather a pernicious and soul-searing tyranny. I did not then accept the then-popular analogy of Munich and Vietnam. Rather, after my summer of 1966, I saw Vietnam through the spectacles of what I had observed in the USSR.

As the eastern European bloc changed so dramatically in 1989 and 1990, and as the Soviet Union itself underwent dramatic changes under Gorbachev, I thought about the students in the Soviet Union a quarter-century before. Would that Soviet professor still order me to shut up, now that the Soviet Union has admitted it committed the massacre at Katyn Forest? Now that Pasternak and Solzhenitsyn can be read in Moscow? Now that the Cold War is thought to be over? Now that the bankruptcy and tyranny of Leninist socialism are universally acknowledged? I do not know. But I do know that I went to the Soviet Union in July 1966 as a liberal Democrat. I returned as a conservative Republican. In 1968, I voted in my first presidential election. I voted absentee from Fort Dix, New Jersey. And I voted for Richard M. Nixon, the Republican, and not for Hubert Humphrey, the Democrat. Just eight years before, in 1960, I had been an ardent John F. Kennedy supporter, wondering how anyone could support his rival that year for the presidency—Richard M. Nixon. I had learned the truth of Stephen Decatur's famous toast in 1816: "Our country! In her intercourse with foreign nations, may she always be in the right; but our country, right or wrong."

I had generally supported American involvement in Vietnam until 1966. Now I strongly supported it. I became increasingly distressed by students my age who seemingly had no sense of duty, no idea of obligation, no concept of responsibility. But many of the students I knew, while strongly opposing the war in Vietnam, did embrace a conscientious commitment to non-violence. To argue, as some did, that everyone who opposed the war was a coward or a "commie," was patent nonsense and did a great disservice to some young men whose moral judgment would not permit them to don a uniform. My college roommate for my last three years at St. A's, Henry I. "Buzz" Talmadge, a self-proclaimed Marxist (although a rather sedate one), took a job as a venereal disease checker in New York City because it was draft exempt. He served, later, as an usher at my wedding, a long-haired hippie in attendance at the marriage of a man wearing the dress blues of an Army infantry officer. We had little to say to each other by that time. My best man at that wedding had taken an engineering position in Connecticut, a draft-exempt job he really did not want; he has it still. Another usher at my wedding, now a federal judge in Massachusetts, managed to get into the National Guard. After his introductory training, he returned to the Commonwealth—and to law school.

We used to have some extended conversations—we the children of the Class of 1968—about our fathers in World War II, or our older brothers, perhaps, or cousins in Korea. St. Anselm's kids—like those in state universities or in other small colleges—tended to be children of the middle class and of blue-collar parents. Most were, truth be told, knee-jerk patriots, some of whom for the first time were beginning to realize that their government was not always and necessarily right. That did not mean, however, that it was always and necessarily wrong—a conclusion incorrectly being drawn by others our age. It did mean that each of us had to wrestle with his own demons to make a judgment about the war and our part (if any) in it.

In the movie *The Graduate*, Ben (Dustin Hoffman) finished college only to wrestle with Mrs. Robinson—not with the draft! Those college students not able somehow to avert conscription (following the end of our cherished 2-S status after graduation) had to confront reality: the U.S. military. Known as a supporter of the war, I picked up the nickname "Jingo"—a superpatriot. Unable to counteract the sobriquet—it did have a nice alliterative quality, "Jingo Jim" Toner—I played up to it. One night, for example, a friend of mine in the college snack bar began playing Donovan's plaintive song of protest about "The Universal Soldier" on the juke box. Unable to stop myself, I punched in Staff Sergeant Barry Sadler's "The Ballad of the Green Berets." We had a duel between these two songs until one of us ran out of quarters, which took a very long time, for each of us had our cheering supporters in short order.

It was in this connection that I decided that I should join the military. I would probably be drafted anyway, so I might as well choose my time and service. Both were restricted. The Navy, Air

Force, and Coast Guard were full of officer candidates, and I intended, after struggling my way through to my degree, to be an officer. I could choose between the Army and Marine Corps. I chose the Army for a very simple reason: after OCS, the active duty time in the Army was two years; in the Marines, it was three years. A dogface, therefore, I would be.

I passed the Army OCS test and interview, and I was offered an OCS enlistment contract. I was to join on 15 May 1958 and, 120 days later, enter on active duty. The recruiter was a bit puzzled when I told him I wanted to take the oath on 14 May instead of 15 May. "What's the difference?" he asked.

"Sergeant," I replied, "May fourteenth is my birthday." On Tuesday, 14 May 1968, "Jingo Jim" joined the Army. Although it had been a very serious decision, based upon a great deal of thought and introspection, my friends and I did have a good laugh. Four months later the joke was on me.

II

On 9 September 1968, after a summer of work, slow-pitch softball, and reading, and having gotten engaged (not necessarily listed in order of importance), I was driven by my parents from Monson, Massachusetts to Manchester, New Hampshire to enter upon active duty. The next day, we recruits were driven to Boston; we took a plane to New York, and from there we took a bus to Fort Dix, New Jersey. Basic Combat Training—or what the Army calls Initial Entry Training—began.[3]

Before that great American rite of passage known as basic starts, however, Army recruits are "received"; that is, they go to the "Reception Center [or Station]," where their initial administrative processing is accomplished. Here a word is in order about initial military processing in the United States. Few people challenge the Marine Corps claim that their boot camp is the most difficult and stressful of the armed forces training given to raw recruits. The traditional Marine Corps task, after all, is storming beaches—something not routinely done by even the most adventuresome teenagers on the block. The Marines, therefore, *must* condition their troops, physically and psychologically, in a way that, say, the U.S. Air Force need not do. Air Force basic training is, somewhat understandably, considered to be the easiest of the four major services' recruit training. At present Air Force basic (conducted at Lackland AFB in

Texas) is only six weeks long, and Air Force T.I.'s (technical instructors) do not have the fearful reputation of D.I.'s (drill instructors) in the Marine Corps or of drill sergeants in the Army. Navy boot camp is not considered by most to be "easy" or "fun," but it is probably not as physically grueling as Army BCT.

Moreover, Army BCT, unlike the training conducted by the Marines, has customarily been given to a very wide variety of recruits. Marine boot camp (except for the relatively rare times when the USMC has accepted draftees—as in the Vietnam War) was designed for the very eager volunteer. Army drill sergeants, and their sister service comrades, are hand-picked and carefully trained non-commissioned officers. In about eight weeks of Army BCT, new soldiers (many of whom used to be draftees) must learn drill and ceremonies, first aid, military customs and courtesies, basic rifle marksmanship, hand grenades, fire and movement tactics, the rudiments of land navigation and map reading, military justice, and a number of other elementary soldierly skills—complemented by what, for many of today's young Americans, is extremely vigorous physical conditioning. Today's Army and Marine Corps recruits are not allowed to use tobacco products; and, for many, eating three square meals a day without sweets and junk food is a new experience—as are leather boots in place of the ubiquitous tennis shoes worn by most teenagers. The premier task of BCT remains the same now as it was more than two centuries ago, when Baron von Steuben drilled Americans during the Revolutionary War: to turn citizens into soldiers.

Intellectually, I knew all that as I arrived at Fort Dix that day in September. (In the Army, basic trainees often get to basic at night; in the Marine Corps, they always get to boot camp after midnight. I can leave to your imagination the reasons for that.) Physically, however, I was somewhat unprepared for our late arrival at Dix, followed by our being marched to a warehouse, where we were issued bedding and where a snarling non-commissioned officer (NCO) or sergeant could not resist asking me if I was a college boy. I think I was smart enough then, or too tired, to offer much of an answer—when in doubt, say nothing! We were allowed a few hours sleep, then awakened at the Reception Center by an NCO who ordered us to march (a new thing to most of us) to another warehouse where we received a vast issue of boots, fatigue uniforms, underwear, belts, caps, jackets, and a number of other things. We were then instructed to mail home our civilian clothes, making our escape efforts, I guess, all that much more difficult. Soon we received our first Army haircut. A barber ran a clipper over our heads a few times, obliterating our hair. If that wasn't bad enough, we then had to pay for it!

Someone told us that, if anyone burst into the barracks and instructed us to "place your wallets on your bunks and get outside for inspection," we should not do it. A number of wallets had been pilfered recently by some inventive person employing precisely this technique. That was my introduction, however unfortunate, to disobeying illegal orders. I was told, too, that I was on KP (kitchen police) that night. That did not seem to be too bad to me—until I learned that it was a twelve-hour shift, from midnight until noon. Recruits at Dix were blended in with soldiers returning from Korea, Germany, and Vietnam for out-processing from the Army. These grizzled veterans roasted the recruits mercilessly. In baseball, the player who hears too well the jibes tossed at him by his opponents is said to have "rabbit ears." I tried not to hear the taunts and jeers of the veterans, but it is hard to be a rookie wherever you are.

Twelve hours of KP! I got the pots and pans, huge containers requiring careful scrubbing with hot water and bristling pads. After about three hours, my hands were bleeding into the water, and I was exhausted. I did not think that I would be shifted, immediately, from being a recruit to teaching strategy at the Army War College, but, somehow, I did not think my Army experience would begin—and I thought that night *end*—with bloody pots and pans. After that KP session, I thought that, at least, I could have waited until May fifteenth (not my birthday!) to join the Army. Lugubriously, I reflected on what the "U.S. Army" over my left pocket meant: "Uncle Sam Ain't Released Me Yet."

Back at the Reception Center, I had a kind of fear of guys who had had some ROTC in college or, worse, of those few guys who were, in Army argot, "prior service." They knew too much; they threatened us raw recruits. They even knew where to tuck in those long, long boot laces. We began to deal with blisters, the recruit's silent enemy. Some of the recruits were unaccustomed to leather footwear, a problem exacerbated today by ever-present canvas shoes; but even in 1968 it was a problem for trainees to wear leather boots without getting multiple blisters.

I tried to have a good, positive attitude, even learning to use military time (1800 is, for example, 6:00 P.M.) and the military calendar system (12 April rather than April 12). But as the few days at the Reception Center passed, I was beginning to wonder whether I had stumbled into some nightmare version of summer camp. As the song has said, we had only just begun.

III

One day it was announced to us that our basic training company had been chosen and our cycle of training time was about to begin. We had to load up on trucks, therefore, for the drive to Bravo Company, Sixth Battalion, Third Infantry Training Brigade (or, as we knew it, B-6-3). We arrived to find a congregation of screaming NCOs awaiting us. Our initial drill was to drop our heavily laden duffle bags and roll peanuts with our noses from where we were toward the barracks, a couple of hundred yards away. This was not a party game I particularly enjoyed, but I decided that I could play it this one time. Soon we were standing in formation, being screamed at for breathing, when I learned that one of my fellow trainees had committed a heinous offense. He had a pen in the left pocket of his fatigue shirt. We quickly learned that only authorized decorations were permitted in the space above the printed "U.S. Army" on the shirt. But on the way to learning that invaluable lesson, we discovered the pecking order at B-6-3.

"Were you awarded that pen?" asked the NCO. The recruit responded with something like, "What, sir?" That seemed polite and innocuous enough to me. It seemed, however, to enrage the sergeant. In a line I swear must be memorized by every non-commissioned officer in the United States Army (because I have heard it so often), the sergeant said, "Don't ever call me 'Sir.' I work for a living."

"Yes, sir, sergeant," said the recruit—to the muffled chuckles of the rest of us. We then learned something called the "front leaning rest position." I could never understand why it was called a "rest position," for, to me, it was the push-up position, which I have never found to be particularly restful. We were to knock out push-ups until our recruit buddy could respond properly to the menacing NCO.

"I am your drill sergeant, and don't you EVER forget it," the very big NCO bellowed. I knew that I understood that. I hoped that my "buddy" would, too, because I was about exhausted from doing push-ups by this time. He did: "Yes, drill sergeant," he responded. Well, we knew who we were, and we knew who he was. Lesson one was over.

That little exercise brought home a point to me: There is some utility in mass punishment. Of course, mass punishment can be used sadistically. But used properly, mass punishment—and mass reward—can be made to serve a valuable end in building teamwork. All of us in the formation that day really wanted our new buddy to answer the drill sergeant properly, for his benefit and for our own. We began to work for one another, to expect proper performance from one another, and, at times, to demand appropriate conduct from one another. Such lessons, properly taught and well learned, are the basis for building any team, whether it is a fire team in the Army or a baseball team in high school.

Some years ago, Peter Barnes published a scathing book[4] in which he contended that basic training "serves as a kind of government-run incubator for what has come to be known as the silent majority." In fact, quite the opposite is true, as Adam Yarmolinsky testified in *The Military Establishment: Its Impacts on American Society*: "basic training [has] apparently served as a form of prophylaxis against authoritarianism."[5] Another critic of Army basic, Peter Tauber, offered a damning indictment in his assertion that one drill sergeant he knew was unfamiliar with the word *eschaton*.[6] I assume that my senior drill sergeant (Sergeant First Class Wilkins) would not have known that word, either. But Wilkins's job was not to teach theology or philosophy; it was to prepare men off the block for combat. Senior Drill Sergeant Floyd Wilkins was an academically ignorant, uneducated man. He invariably referred to us as "mens." As a drill sergeant, however, this gargantuan, stentorian black man—he must have been 6'4", 230 pounds—was exceedingly competent, fair, organized, and enthusiastic. He knew what he was about. Judged from the standard of humane, liberal arts, basic combat training is a degrading and dehumanizing eight weeks. But its graduates, or at least some of them, go on to fight Germans, Japanese, North Koreans, North Vietnamese, or Iraqis—none of whom ever ran away when a soldier screamed "eschaton" at them. In a very real sense, although Wilkins had no education, he was schooled. Tauber, although he had completed college, had never been "schooled." Some people give education a bad name.

Individuality is a prized possession of our democratic country. In the military, however, it can be pernicious. The BCT of 1968 taught teamwork in odd ways. There is a story that Pope John XXIII, examining blueprints of a Vatican building, observed, "Angeli non sumus" ("We are not angels"). This puzzled the builders until they noticed that the building had no toilets! In the BCT of yesteryear, there were toilets—a cluster of five commodes in a semi-circle. Now *that* was a culture shock! So were my trainee buddies—from all over the United States. At the time, the trainees were ER, NG, US, or RA. The Regular Army guys took the worst razzing: "You mean, you *joined*?" we would be taunted. My number (RA 11750662) gave me away. (Today, social security numbers are used.) The National Guard and Enlisted Reserve guys got to go home after basic and AIT (about which more follows). The US guys were the draftees—in for two and go. Every day in made them "shorter"—closer to ETS (estimated termination of service). The word s*hort* came to refer to people with decreasing time of service, those nearly "out." Two old friends of mine from Monson, Massachusetts, having learned from the post locator where I was, stopped by to see me one Sunday afternoon. They were both in the Guard and about to go home. (Their AIT had been at Fort Dix, too.) They meant well and I appreciated the visit, but knowing that they were on the way home did not cheer me up much.

Soon after arriving at B-6-3, I had my first meal in the mess hall. The sign said, "Take all you want. Eat all you take." I took a lot. I did not eat it all. An NCO staking out the return-tray area noticed my delict and ordered me back to the table to consume what I had left on the plate. I didn't eat it because I didn't want it. "Then why did you take it?" I could hear my Uncle Sam asking. I managed to get the garbage into a couple of napkins and, holding the mess in them (concealed under my tray) and displaying a clean plate, I walked to the return-tray area again. The NCO grunted and let me pass. After, I reflected that I had just bet my life on that trick. Had I been caught, I am sure that I would have learned very quickly how to eat both garbage and napkins. At least I had won. I thought that the Army should give me an award for heroism—maybe a pen I could carry in my left front fatigue shirt pocket!

Another time I was in the mess hall I was not quite so lucky, and I lost a lot of face with my trainee buddies. It was breakfast time, and I was hungry. I loaded up on cream of wheat, always among my favorite breakfast cereals. When I poured on the milk and sugar, I noticed some people looking at me. It was definitely the worst cream of wheat I had ever had. Little did I know it was grits. We didn't have grits in New England. After eating as much of the mess as I could stand (there was too much and it was too messy to try the napkin trick again), I decided that the real reason for the Civil War was that the north was afraid that the south could somehow establish a national diet with grits included. The Army never taught me table manners. But I watch what I eat.

Most of the NCOs in basic were, on balance, pretty good. But I had learned the great truth of Lord Acton's dictum ("Power tends to corrupt; absolute power corrupts absolutely") at the Reception Center. An E-2 (at that time, they wore one yellow chevron—a "mosquito wing"—on their fatigues) was supposed to march us from X to Y to do something. He had completed only basic training but, to me, he was equal in rank at least to General Westmoreland! Power went to his head. From somewhere he produced a razor and had some of us dry shave; he had us do grass-drill (a series of calisthenics very quickly done); he had us run. I would, even today, like to find him. And kill him. More seriously, one realizes that it is exactly that kind of will-to-power complex from which leaders must be free if they are to exercise their responsibilities wisely and fairly. I may not have been the best officer in the U.S. Army; but I learned one thing from that E-2: I never ordered any of my men to dry shave!

Among the things basic training must teach is, in a nutshell, that "you're not a kid any more." You can die for your country. You can kill for your country. Some young men grow up a great deal in, and perhaps because of, basic; some don't. In the Army of today, as previously alluded to, tobacco products are forbidden so that the smoking habit can be broken; even then, some will be recidivists and return to smoking. But basic is supposed to help young men and women assume responsibility and respond intelligently to authority. This is brought home through something known as BRM (basic rifle marksmanship) and grenade throwing. Some new soldiers have handled firearms for years; to others, the M-16 is entirely foreign. The Army teaches marksmanship extremely well to trainees from an infinite variety of backgrounds. To college graduates, rifle instruction appears feckless—almost dumb. Yet in the course of teaching rifle marksmanship to so many hundreds of thousands of trainees, the Army drill sergeants have learned some remarkable lessons. Some of the smartest trainees, silently laughing at the "dumb drills" (as the drill sergeants are sometimes covertly called), fail to understand that the instruction offered reaches a wide variety of trainees.

As an example of that, we were taught to break down our rifles to clean and oil them and to be able to check them rapidly, even in the dark, if they jammed. I thought the process was pretty simple, and I did not pay very close attention. Some of the trainees, although perhaps high school dropouts, were able to field strip their rifles much faster and more efficiently than I, and I thereby learned a lesson not related to BRM: some of the most uneducated troops could prove to be among the most intelligent in certain chores. There are different kinds of intelligence. A little humility can be a good thing. And doing things the Army way isn't always a bad idea.

There is a story about basic training at Fort Benning that reinforces all this. During BCT, trainees learn to throw grenades. I always hoped that my trainee buddies had played at least a little baseball! The standard instruction to trainees on the grenade range are these: if you pull the pin on a grenade and drop it, do not move; assume the position of attention, and allow your drill sergeant to retrieve and to throw the grenade. One afternoon at Benning, a trainee dropped his grenade after pulling its pin; he bent over to pick it up—and butted heads with the drill sergeant in the pit with him. The same comedy happened on his second effort. Then tragedy: the grenade went off, killing the NCO and badly wounding the trainee. When I went through airborne school, a long time after basic, I remembered that Army training, although often slow and tedious, is remarkably effective—unless you're just too smart for it.

Of course, you can also be too dumb for it. Basic trainees are not taught to work with mortars. But if a mortar round "hangs fire" and doesn't pop out of the tube, the immediate action drill to rectify the situation is, really, to kick the tube and thus "encourage" the round to cook off. Army training actually does tell the trainee not to look into or down the mortar tube! But I am aware of one episode, again at the Infantry School at Fort Benning, where a trainee did look down the mortar tube just as the round cooked off. Decapitated. The Army can protect a person from himself only so far.

IV

Much ink has been spilled about the cruelty of bayonet training and of hand-to-hand combat training. The manual combat skills I was taught in BCT were designed to save my life. I know personally one former officer who says he owes his life to the conditioning and combat skills the Army taught him.[7] Many people are offended by manual combat. The Marine Corps today, for example, teaches its boots or trainees a "nervous system overload" concept in which the young Marine is taught to break the elbow or knee of his opponent and then to hang onto the ruptured limb if at all possible. This creates such incredible pain in his distracted opponent that the Marine can easily kill him. Army Rangers learn a wide variety of techniques for silent killing in order to carry out the kind of patrolling that they are justly famous for. Although Army basic of today has little hand-to-hand combat, the BCT of 1968 had a great deal of it. I was taught the "rear-strangle-takedown hold," for example, in which a sentry can be dispatched quickly and quietly. We were taught the basics of hand-to-hand combat: go for the eyes and the gonads. These rules do not easily square with the protocol of most garden parties.

Neither do pugil sticks. Soldiers are still taught the basics of the bayonet with a long, padded stick. After learning the basics of bayonet fighting, trainees practice—on one another. Wearing pads and face masks, they pound one another with the pugil sticks. The Marine Corps uses the sticks, also, even pioneering in the use of a sandbagged fighting pit in which pugil-stick antagonists bang it out, under the watchful eyes of the D.I.'s. Soldiers and Marines are not being prepared for upscale partying. They are being readied to kill, or to risk their own lives, as agents of the American government. Training made too easy or training made too unrealistic will kill Americans when they do get into actual combat.[8]

In late 1990 and in 1991, President Bush ordered thousands of Army National Guardsmen and Army Reservists to active duty in the Persian Gulf. Thousands of these soldiers had not received any actual combat training beyond what they learned in basic. Every soldier and every Marine must be prepared to assume the role of a combat infantryman. Korea taught us that lesson. If a position is overrun by the enemy, that same enemy is not interested in an American soldier's MOS. Young trainees on the firing line of Army ranges, in grenade pits, and in bayonet training are being prepared to accomplish their mission and to protect their own lives. Army trainees are taught, in bayonet practice, that the "spirit of the bayonet" is to "Kill!"—which they are instructed to scream as they plunge the knife deep into straw men. It all seems bitter and brutal. But to impart lessons of timidity to soldiers, and then expect them to act as soldiers on the battlefield would, first, effectively prevent them from doing their military duty and, second, virtually sign their death warrants. More along these lines will appear in the last chapter of this book.

Much the same can be said about CBR/NBC training. Chemical-Biological-Radiological training or Nuclear-Biological Chemical training seems horrible to many. Yet every graduate of BCT has undergone it. In essence, aside from some classroom training, it amounts to little more than walking through a Quonset hut and removing the protective mask so that the trainee can get a whiff of tear gas (CN or CS). This induces some tearing and a prickly feeling in sweat glands; first aid is just letting fresh air get at the gas (really crystals or particles) and not rubbing at it. The discomfort is usually minor and the effects not persistent. This introductory training could save thousands of lives in a chemical war. But the soldiers must learn some self-discipline and they must endure some discomfort. The tendency, for example, to rub at the crystals (only pushing them deeper into the skin) must not be given in to. The spirit of immediate gratification, should it prevail, would only make a small discomfort a major one.

Another example of the "discomfort lesson" is the barracks "fire watch" routine. Trainees have to pull guard duty watching out for fires. This is a great inconvenience when you do not get a great deal of sleep and your watch is, say, from 0100 to 0200. (It's not so bad if your watch is right after lights go out or just before reveille.) You learn the lesson that you must take your turn; you must sacrifice for the group. There clearly must be a group mentality—and ethic—at work. In the Army of the 1960s and 1970s, soldiers had to appreciate the fact that there were no blacks and there were no yellows and there were no whites. It was one green Army. I learned that lesson, and I was directly challenged on it later, during my tour of duty as an officer, an episode to be related later in the book. Shared burdens and discomforts add to the psychological problems of the new soldier as he begins to learn that, whatever his background and interests and education, he has much in common with all the other soldiers in his group, however different they may be: survival and success. It is a lesson all trainees must learn. Thus the Army truly begins to become an *American* fighting unit. In airborne school, for example, when you check your buddy's equipment (and he checks yours), you do not check his color—unless you are a complete scoundrel.

Marching and doing field training exercises in rain, snow, sleet, mud, very hot or very cold weather increases, oddly enough, the camaraderie of most soldiers who are going through these things together. Freshman recruits at VMI, The Citadel, North Georgia, and Norwich—and at all the federal service academies (as we will explore in the chapter on ROTC)—have learned this lesson, together, for decades as rats or knobs or rooks or frogs (or plebes). The perceptive trainees, too, realize, finally, that their drills are up before they are and that they are marching right along beside them. It's a practical lesson in leadership that trainees can and should recall throughout their military and civilian careers.

V

Army training, especially in Infantry OCS (which I went through) for six months (in 1969) and Ranger School (which I did not go through) for eight weeks, is intent upon teaching, too, the need to understand our own physical limits. Basic training gives a small taste of the same thing, as does Infantry AIT. Some of the trainees are good athletes in excellent physical condition. (To listen to drill sergeants and drill instructors at today's Army posts and Marine Corps bases, however, they are rare!) It is important even for them to realize that military duty can be exhausting and that sometimes they may have to compose a five-paragraph field order to a patrol when they are physically and mentally exhausted. So trainees are pushed hard in order for them to learn two lessons: the limits of their physical abilities and the power of their own mental toughness to overcome and extend those limits when necessary.

In basic, this lesson about physical limits could quickly be taught. When a trainee made an error, he (and some of his buddies) might be made to stand at attention, holding their rifles up and parallel to the ground, arms outstretched in front of them, palms to the ground. Even the superbly conditioned athlete will soon find his muscles rebelling, twitching and jerking mightily, against the strain. That little drill was not something I would inflict on others; but it revealed how quickly our vaunted physical powers can be exhausted. Wise soldiers learn to know both their own limits as well as how (and when) to extend those limits. At some times when the individual, left to his own devices, would quit, the soldier, with responsibility to his unit, cannot. The soldier must learn to reach deep within himself to know, and to do, what is right for as long as possible.

Of course, trainees must learn to deal with stress. Some of the stress is minor, such as the need (at B-6-3) to complete a certain number of pull-ups on a hang-bar outside the mess hall before eating. If a trainee could not perform the magic number of pull-ups, he could still eat—but he took quite a going-over by the drills. I didn't mind doing the pull-ups, but I did want to wash my hands before eating—something we were not allowed to do. Some of the stress was even mildly humorous. We received our shots from "jet guns," which actually blow the serum into the bloodstream through the pores, provided you do not move while you're being injected. Flinch, and you get a tear in your skin. Some guys always flinch! Those guys, bleeding profusely from what amounts to deep (but normally not dangerous) cuts in their arms, are always sent back through the lines for others to see. "Better pay attention, mens!" yell the drills. Some of the stress is provided by the trainees themselves. We lived in an old platoon-bay type of barracks, two squads upstairs, two down. There were no dormitory-like arrangements. You lived with your fellow trainees, all together. (Parris Island and, I believe, San Diego preserve this tradition in the USMC today; the Army, which has largely gone to college dormitory-like barracks, should return to it.) If one trainee doesn't have high personal hygiene, he is normally strongly encouraged by his buddies to modify his personal cleanliness regimen to be more in keeping with their preferences. One of our B-6-3 buddies was not particularly clean, a fact brought home quickly to all of us by our noses. He didn't shower. One night several trainees (and I hasten, truthfully, to point out that I was not among them—but neither did I go to his aid) jumped the offending "buddy." They forced him into the shower, where he was involuntarily lathered up and rubbed all over, while naked, with a G.I. brush. This attracted his attention, and he thereafter was a model of personal cleanliness.

At one point in BCT, I received an alumni publication from St. Anselm's. I read there that an old friend, First Lieutenant Jim Said, had been wounded in action in Vietnam. Said had graduated two years ahead of me; he had gone through BCT, and AIT, and OCS, too. Now he was a casualty in Vietnam. Somehow the full meaning of the training came home to me. It was not a joke. Said recovered and I later bumped into him at Fort Benning.

After seven weeks of training, I thought I was learning my lessons fairly well. In the end, however, I never graduated from basic training. For reasons no longer entirely clear to me today I decided to take advantage of free Army dentistry and have my wisdom teeth extracted while I was in Basic. Bad judgment! At the same time I was suffering from the effects of those extractions, I contracted what the Army calls URI—upper respiratory infection. In my case, that amounted to pneumonia. I made my way to a dispensary, coughing and spitting blood (from my gums, not from my lungs)—not exactly combat ready. I was immediately sent to Walson Army hospital at Dix. "I can't go there," I pleaded, "because I'm supposed to graduate from basic in a week." "Oh, God," I thought, "I'll be re-cycled to Week One." I got even sicker from the thought.

Soon I was so sick I no longer cared. It took me about ten days to get out of the hospital. The good news was that I didn't have to be re-cycled, for I had done fairly well in basic. The bad news was that I had to pass a test on my basic skills. At once. That didn't bother me too much, except for the Physical Fitness Test. That bothered me a lot. At the time, that meant I had to pass a group of physical fitness tests including a horizontal ladder hand-walk; a run-dodge-and-jump exercise; a soldier carry; a grenade throw; and a one-mile run in boots. (Today, the Army's PT test has push-ups, sit-ups, and a two-mile run—in sneakers.)

Normally, the run didn't bother me, but it was November; it was cold; I was still weak and a little sick. The NCO was a big help, telling me that if I didn't pass the PT test, I *would* be re-cycled. I made the run. Barely. But I never graduated from Basic Training. Not officially anyway. I was sent to a holding company while orders were prepared for me to go to AIT (Advanced Individual Training) at Fort Dix, for eight weeks, in infantry—basic combat training, only advanced. Again I thought about the "wisdom" of having joined the Army on my birthday!

Although I still have my BCT yearbook, I have seen, in all the years since I got through BCT, only one or two of the guys I went through that experience with. Some were killed in Vietnam. Some perhaps became heroes. Maybe one or two even wrote about it somewhere. Neil Simon's play *Biloxi Blues* tells the story of one Army BCT company. Simon contends that he liked the Army for the best of reasons: he was young. But his group disbanded before it had to go overseas. World War II ended. For my BCT company, the war was just beginning—as was my training.

VI

In the U.S. Army of 1968, the graduate of basic combat training proceeded, whether he was ER, NG, RA, or US, to what was called his MOS (military occupational specialty) or skill training. Training was provided to soldiers in scores of skills from accountancy to zoology (well, all right, doghandler). In today's Army, soldiers who have chosen infantry training receive one-station training at Fort Benning, Georgia. In 1968, however, soldiers could take basic at a number of Army posts and then be shipped elsewhere for their

Advanced Individual Training—as, of course, can still happen today. In my case, I stayed at Fort Dix for training in my chosen field of infantry.

I chose infantry skill training because I had earlier signed a contract guaranteeing me Officer Candidate School. In turn, I had chosen Infantry OCS at Fort Benning. To many, that would not appear to be the most desirable choice! I was never a gung-ho infantry type but, when I joined the Army, I had a choice of attending one of three Army Officer Candidate Schools then in operation. One was at Fort Belvoir, Virginia; it was principally for engineers. One of the chief reasons I had chosen to go to Saint Anselm's College was because it did not require mathematics. I thought that, if I were going to attend OCS, I wanted to graduate. Engineer OCS was out! I could have attended OCS at Fort Sill, Oklahoma. But I knew that there was a fair amount of math in Artillery OCS there; it, too, was out. That left Fort Benning, Georgia, home of the Infantry, the Queen of Battle. I figured that Infantry OCS would be very physically demanding; but I thought I could handle the math there (believing there wouldn't be any!). I chose my fate, therefore, in deciding to attend the U.S. Army Infantry School in Columbus, Georgia—principally because I was, as the term from a recent book suggests, an "innumerate," untutored in mathematics.

But first, I had to complete my week at the Reception Center, followed by my eight weeks of basic. After I had struggled my way through those, I faced another eight weeks of AIT in infantry, still at Dix. After that, I faced 23 weeks (six months) of OCS at Fort Benning.[9] I wanted to be an officer, and I was beginning to think that my decision in high school not to attend any college with ROTC was about as dumb as joining the Army on, of all days, my birthday. If I had been in ROTC, I would have pinned on the gold bars of a second lieutenant on my college graduation day.

My AIT assignment was to A Company, First Battalion, First Infantry Training Brigade, or A-1-1 for short. At least the barracks looked better. They were more modern, albeit a dull gray color, buildings in a different part of the post from B-6-3. (A word about that: In the Army, a military installation [with a very few exceptions, to complicate matters] is a post, not a base. *Base* is a term reserved for Air Force, Marine, and Navy installations. If this seems picky, ask a Marine what he normally stands on [even on dry land], and he will reply, the "deck" when the rest of the world, to include Navy people, will say the floor.) By now it was November, and it was getting cold.

That, in itself, did not disturb me until I remembered that the purpose of Infantry AIT was to prepare RA and US graduates for combat in Vietnam, where it was very hot, almost immediately following graduation.

During the second day at A-1-1, I was standing in the noon mess line (at parade rest, of course) when, from somewhere behind me, I heard this unsettling sound and greeting: "Pssssst, Jingo! PSSSSST, JINGO!" The last thing I wanted to do was to turn around and look; one did not do such a thing from the position of parade rest. There were still NCOs in AIT; they still wore the Smokey the Bear campaign hats that drill sergeants wear (but we didn't have to call our AIT NCOs "drill sergeant"); and I did not want to be the one to set the push-up example for my new company. A third and imperious "PSSSSSSSSST, JINGO!!" convinced me, however, to turn around, furtively, and find out who this menace was. There, standing at parade rest, grinning a cheshire cat grin, was my old St. Anselm's buddy, Doug Jhoram. It was the continuation of a long relationship. Jhoram and I were to go through AIT, OCS, and Airborne School together. He went on to Ranger School and to a subsequent Army career; I went home to be married after Airborne School and subsequently entered graduate school after my service was concluded. I was glad to see Jhoram that day—but happier still that our NCOs hadn't seen us exchange tacit greetings.

I was in decent shape after basic training, so the physical side of AIT didn't bother me too much. We did a great deal of marching and running, but much of our training was of the classroom variety: tactics, map reading, first aid, weapons orientation, and the like. After the classroom lectures and demonstrations, we would hike out to the ranges for practical exercises. There were rumors that the training, especially in anti-tank weaponry and in explosives, was substandard; I had no basis by which to judge. One day, however, I learned a lesson that I hope I have never forgotten.

We still had to do KP, based on our position on a roster. The first time I had KP, the guys in A-1-1 were learning about the LAW (light, anti-tank weapon). I was washing dishes, using my hard-won Reception Center training. I knew that, later, I would be tested on the LAW so, when I had some free time later, I went to the second lieutenant who was the executive officer of our AIT company and respectfully asked him for help on the basics of the LAW. That weapon is exceedingly easy to learn to use (I know now). He knew nothing about it; nor did he tell me to come back at a later time to go over it.

He summarily dismissed me. I really didn't mind that he couldn't drill me on the LAW; I didn't expect him to know everything. But I did object strongly (to myself, quietly) when he simply sent me away. I actually thought I was being a pretty good soldier in trying to find out about this weapon, which I knew I would see again in OCS.

About two years after that, I was sitting in my office in Germany when an NCO came in to see me, asking me (as an infantry officer) to show him a few things about the .45-caliber pistol. Some of the things he asked me, I knew; others, I did not. I asked him if he would please come back tomorrow, and we would go over it in detail. I then went to the company armorer (weapons expert), refreshed my knowledge of the .45, and prepared in minute detail for the meeting the next day with the NCO. Of course, I was able to answer his questions, with specific and abundant knowledge. He seemed impressed with my knowledge, which, I will admit, was a bit meretricious; but he seemed more impressed with my attitude and willingness to teach him whatever he wanted to know. We got along very well, and I attribute that to what that AIT second lieutenant had "taught" me. Some of the lessons we learn are taught for the best when, paradoxically, they are taught for the worst! I should say that I had been honest with my NCO. I told him frankly that I did not know some of the things he wanted to know, but that I could find out—and would, at once. My AIT second lieutenant could have said the same thing to me, and I would have respected him. As it turned out, he was only a "Lieutenant Fuzz," something I resolved on the spot never to be.[10]

There were two lieutenants in the company. The other one was a first lieutenant just back from a tour in Korea, where he had been sent after OCS. The company commander was a captain, but I can recall nothing about him, which may be a commentary in itself. (At least I can recall my commander in basic, who was a first lieutenant whom I didn't like!) The lieutenant who couldn't, and wouldn't, explain the LAW to me was an ROTC graduate of a public midwestern university. The more experienced lieutenant was a Benning product. At A-1-1 there were a number of us headed for Benning OCS, and the first lieutenant talked to us about what lay ahead. He didn't seek to hide his feelings about ROTC lieutenants. "They have it too easy," he said. He told us that Infantry OCS was the hardest six months he could ever have imagined. He said, "You don't sleep. You don't eat." I figured that was an exaggeration. I hoped. Still, I began that day to take vitamin pills, and I didn't quit, as it turned out, until after Airborne School.

This chat with "Lieutenant Korea" was my introduction to the tension that exists in the Army (and, to some extent, in the other services) about "source of commission." I was to learn later that ROTC graduates whose commissions were in infantry attended IOBC at Benning (the Infantry Officer Basic Course). They got to drink coffee and eat doughnuts on the breaks during classes in Infantry Hall. We in OCS used to stand at parade rest in the halls during breaks. IOBC types used to shine their boots with "Hershey Bars" (they didn't gleam), whereas OCS guys used to have to keep spit-shined boots. OCS guys had to maintain starched, pressed, fresh fatigues. The IOBC types seemed to have a corner on "oiled" and rumpled fatigues. We wore shined helmet liners; they wore the greasy baseball caps of the day. We hated IOBC types. Lieutenant "Korea" and Lieutenant "Fuzz" persuaded me that there was something to the stereotype. But, as I was to learn later, this wasn't really true. Whether the officer was from OCS, ROTC, West Point, or a direct commission, what mattered, ultimately, was the character of the officer.

VII

One of the problems with the training was that it was conducted in very cold weather, with snow on the ground. We would go on an FTX (field training exercise) and sleep (a little) in a squad hut, arising early in the morning to dig a cathole (a one-foot deep hole for use in personal sanitation), if we could dig at all. Then it was off to morning chow, which amounted to semi-frozen grease looking a little like eggs or sausage. The coffee was always luke-warm. Then we would rinse our metal mess kits in large, steaming buckets (which, after a few troops used them, were full of steaming grease). One of the cooks, a southern draftee, could be heard to repeat endlessly throughout our repasts, "Good trainin', good trainin'." We thought about accidentally killing him. Some of the difficulties we faced while training for combat in Vietnam in snow-covered New Jersey were mildly humorous. Others, we thought, were potentially deadly.

One of our NCOs, Sergeant Edison, was a tough, combat-tested, no-nonsense soldier with little use for officers and with no use at all for non-combat officers, meaning all of ours, even "Lieutenant Korea." But worse than officers, in his view, were readers—especially of a little book then making the rounds of those of us in AIT who were interested in reading. It was John Sack's *M*,[11] a study of a Fort

Dix AIT company that supposedly was decimated in Vietnam after its graduation from the very training that we then were undergoing. Sergeant Edison thought the book a disgrace, although he clearly hadn't read it. At the same time, I didn't want Sergeant Edison to know that I had read it, for I would have lost a lot of face with him if he thought I would stoop so low as to read that "anti-military" book. Sergeant Edison seemed to like and respect me, and one part of me hated to lose that, especially if he thought I was an "intellectual."

Fortunately, I never had that problem. I was resolved, at least, to be honest. If he had asked me if I had read the book, I would have told the truth. But the problem became someone else's. One day we had a surprise inspection. Sergeant Edison did not discover any drugs or alcohol or firearms; but he did discover a copy of Sack's book in the footlocker of a hapless AIT buddy of mine. That did it! That guy received one of the severest tongue-lashings I have ever heard. He was a "commie," a "pinko," a "queer," as well as a number of other things—none of which I had ever suspected about this guy—now unmasked by Sergeant Edison. He then ripped the book up and stormed out of the squad bay. Whether any of this ever reached the company commander, I do not know; but *M* was for sale in the PX right next to A-1-1 (not even the bowdlerized version), and Sergeant Edison, having had his say, never alluded to it again. Neither did the rest of us.

I began to think, however, that Sack's condemnation of the training at Fort Dix AIT was merited. When we qualified with the M-60 (machine gun), the day was miserable. I was realistic enough to know that infantrymen must defeat the weather, as well as the terrain and the enemy. To train exclusively in beautiful, sunny weather would be foolish. But this particular day was a "Dix Deluxe": sleet, rain, hail—and the M-60 range. As I fired the crew-served weapon, an NCO in the tower above me was supposed to be observing downrange, determining the accuracy of my fire. I had to prove myself with this basic infantry weapon. But I couldn't begin to see where I was firing! Visibility was near zero; the rain was making it impossible for me to descry anything in the distance. But the NCO in the tower was commending me for the accuracy of my fire. I decided to test the system. I told my buddy, who was feeding belts of ammunition to the gun, "Watch!" and I began to fire bursts at a considerably elevated, imaginary target. In fact, I had a fear of mistakenly shooting down an Air Force plane en route to McGuire Air Force Base, contiguous to Dix. I continued to receive plaudits from the NCO in the tower. I thought about directing a few rounds his way to see if his evaluation of my gunnery might change, but the fear of serving a few years in the Dix stockade deterred me.

Having been an officer now myself, I no longer hold that NCO responsible. There was a range officer, and he should have had the sense to understand that training on that day was absurd.[12] Had we been under attack, we would have had to fire away as best we could. But a training mission carried out merely to fill in the squares on a report was worse than useless. The brighter trainees began to realize that training here was a joke—even to the officers involved. How, then, could we take it seriously? But it was our lives that were on the line. I was supposed to be great with a machine gun, the reports said. In fact, I received plenty of subsequent training on the M-60 at Fort Benning. But what if that had been the end of my training? What if the next time I fired that weapon had been in support of ground combat operations in Vietnam? That day at Dix the NCO in the tower and the range officer were, in effect, telling my hypothetical Army buddies of the future that "This guy's O.K. He checks out on the 60. He can lay down a base of supporting fire to protect your movement. You can trust him with your lives." They lied that day. I chalked it up as another positive lesson, negatively taught. To this day, even as a teacher, I will not grade a student as satisfactory unless I believe he's satisfactory.

I learned to sleep at AIT with a pillow over my head. I still sleep that way. It would cut down some of the light and noise, for I customarily tried to go to sleep once I had everything done. In contrast to me, some people seemed to have boundless energy. One guy, an AIT buddy from Illinois named Mike Delchamps, was that way: indefatigable. He was a natural leader and I admired him—except that he became too much of a leader. That is, his imperious and importunate attitude, instead of generating respect, began after a time to generate animosity. One of the valuable things about trading off positions of leadership (which the Army has its students do in OCS, in Ranger School, and in ROTC Summer Camp) is that, if you watch closely, you can learn a great deal about how far—and when—people can be pushed, especially by a peer. Yogi Berra is supposed to have said, "You can observe a lot—just by watching!" How true it is. Two of the guys in my AIT platoon were twins, and both wanted to go into Special Forces after OCS. They seemed, perhaps for obvious reasons, caught up solely with themselves, and I have often wondered how they fared as officers. After AIT, I lost track of them.

But I knew I did not want to serve under Mike or the twins. I am not sure that there's an object lesson in this (e.g., trust the man who sleeps with a pillow over his head) except that I learned in AIT that a leader not comfortable with himself will alienate those for whom and to whom he is ostensibly responsible.

I spent Thanksgiving 1968 at A-1-1. Three weeks later I received leave to go home for Christmas. I learned again that I was, to some extent, a pariah from my own generation. As I put on my greens to fly home, I thought of the months of training I had gone through since I entered the Army. In fact, I was proud of my uniform; many of my fellow soldiers, however, said, "As soon as I am out of these gates, off comes the uniform—at least as soon as I can buy a pair of jeans and a sweat shirt." They also planned, of course, to cover their military haircuts. Some planned to buy wigs. I always thought that the U.S. Army uniform symbolized our country, and I was proud to wear it. A comment like that, however, usually engendered the rejoinder that I was a goody-goody, so I usually refrained from the comment in the first place. In my heart, I was still a jingo.

After the Christmas leave, we all returned to Dix—except for one guy who had burned to death over the holidays. I wasn't thrilled with returning to the training at Dix. When I got back to the barracks, I still had a few hours on my leave, so I went to a post theater and saw *The Night They Raided Minsky's*, which I thought was mildly amusing. I was to see that film a couple of years later when I was in Germany; then I thought it was hilarious. I guess when I returned to Dix, I was much more depressed than I knew.

There is a personal footnote to this chapter. In June 1989 my son Patrick was graduated from Northfield Junior-Senior High School in Vermont. The graduation ceremony was held on a Saturday. On Sunday, Patrick, who had joined the Vermont Army National Guard a few months earlier, was to report for processing to Manchester, New Hampshire, from which, the next day, he would make his way to Fort Dix, New Jersey, to take his Basic Combat Training. I had mixed feelings about his decision to join the Army and, of course, to go through basic. Pat was only seventeen—and a very young seventeen at that. When I thought about the rigors he would endure in basic and after, I was deeply saddened. At the same time, I was proud of his willingness to accept the challenge and I hoped that, in time, his basic training experience would redound to his manly benefit. In 1989, twenty-one years after I had made my hesitant way to Fort Dix for

BCT, my son followed suit. Eight weeks later, he flew into Burlington International Airport in Vermont, where his mother, his two brothers, and I met him. He had graduated from BCT.

NOTES

1. The twenty-sixth amendment granting suffrage to eighteen-year-olds was ratified on 1 July 1971.

2. David J. Miller had burned his draft card in New York City on 15 October. He had come to New Hampshire to set up a "peace booth" at Saint Anselm's. Some St. Anselm's students, including "Salmon," began a fast at St. Joseph's Cathedral in Manchester concerning the Vietnam War and, in particular, the Catholic Church's response to the Berrigans. For details, see *Facts on File*, 1965, pp. 370, 383, 454, 480. By early 1968, St. Anselm's had become the vortex of debate during the Democratic primary, in which the incumbent President (Johnson) narrowly defeated Senator Gene McCarthy of Minnesota—in part prompting Johnson's resignation from the nomination battle on the night of 31 March 1968.

3. For more information, see, for example, Jeff Bradley, *A Young Person's Guide to Military Service* (Cambridge, MA: The Harvard Common Press, 1987); Walter J. Jackson, *Soldier's Study Guide* (Harrisburg, PA: Stackpole, 1990); Texe W. Marrs, *You and the Armed Forces* (New York: Arco Publishing, 1983).

4. Peter Barnes, *Pawns: The Plight of the Citizen-Soldier* (New York: Knopf, 1972), p. 84.

5. Adam Yarmolinsky, *The Military Establishment: Its Impacts on American Society* (New York: Harper & Row, 1971), p. 401.

6. This absurdity is taken from Peter Tauber's monumentally naive work *The Sunshine Soldiers* (New York: Simon & Schuster, 1971), p. 12. Tauber detected a lot of the nonsense and stupidity of basic. He is not wrong for calling the reader's attention to such stupidities. As alert, however, as he is to the foolishness of some practices, he seems to have a blind eye for the sense of duty, honor, and country that motivates so many of the people who wear the uniform.

7. See David Donovan, *Once a Warrior King: Memories of an Officer in Vietnam* (New York: McGraw-Hill, 1985), pp. 210–211. This is simply the best book I know about being an officer and a gentleman. "Donovan" is a pseudonym for Dr. Terry Turner, who now teaches in the medical school at the University of Virginia.

8. In this regard, consider Thomas L. Friedman's point that "American statesmen must know how to play hardball" and that "Middle East diplomacy is a contact sport" in *From Beirut to Jerusalem* (New York: Farrar Straus Giroux, 1989), p. 508. The American tendency appears to be, both in soldierly as well as diplomatic skills, too quixotic. In the Army and in diplomacy, the two key axioms to be learned are: "Don't expect too much" and "More sweat now, less blood later." Soldierly and diplomatic training must be realistic, prudent, and demanding.

9. Army OCS is now fourteen weeks long.

10. Those interested in BCT and in leadership should see George C. Wilson, *Mud Soldiers: Life Inside the New American Army* (New York: Scribner's, 1989).

One of Wilson's suggestions is that officers go through basic. Other books of interest in this regard are Larry Ingraham, *The Boys in the Barracks* (Philadelphia, PA: Institute for the Study of Human Issues, 1984) and Charles Moskos, *The American Enlisted Man* (New York: Russell Sage Foundation, 1970), which is still valuable.

11. John Sack, *M* (New York: The New American Library, 1966).

12. For an interesting hypothetical case study about the possible conflicts between training realities and safety considerations, see Anthony E. Hartle, *Moral Issues in Military Decision Making* (Lawrence, KS: University Press of Kansas, 1989), pp. 144–145.

2 PRECEPTORY: INFANTRY OCS AND AIRBORNE SCHOOL (FORT BENNING, GEORGIA)

No man can be a good officer who does not undergo more than those he commands.

—Xenophon, *Cyropaedia*, about 360 B.C.

No pain, no palm; no thorn, no throne; no gall, no glory; no cross, no crown.

—William Penn, 1669

Founded about 1119 in Jerusalem by French knights, the Knights Templars were members of a religious military order. A preceptory was the estate or manor—a community—of the Knights Templars. A *preceptor*, of course, is a teacher or tutor. An officer candidate school is, at its best, a modern preceptory—but one in which the object lesson of General George S. Patton, Jr., of the twentieth and not the twelfth century, is held supreme. In discussing leadership, Patton once demonstrated his point with a china plate and a wet noodle. Patton held the plate perpendicular to a table and tried to push the wet noodle up the slippery, slick surface. His effort, of course, failed.[1] His point was that military units, like a piece of spaghetti, can be led (pulled) only from the front—not pushed from the rear. The preceptory of the American infantry is in Columbus, Georgia, at Fort Benning, home of the U.S. Army Infantry School, and of airborne and ranger training. Suitably enough, the motto of the Infantry School, much as Patton might have taught, is "Follow Me."

Every American armed service has, or has had, officer candidate schools. Over the years, these schools, these modern preceptories, have produced thousands of commissioned officers, but there are very few accounts, factual or fictional, about this vital source of commission. This account, of course, cannot pass as a general representation of OCS, even as it was then, in the United States Army. When I was in OCS (February–August 1969), the course lasted twenty-three weeks. By contrast, during World War II, OCS was about twelve weeks long. (It was customary to talk about "ninety-day wonders.") Army OCS is now about fourteen weeks long. The Marine Corps, Navy, and Coast Guard also have OCS; the Air Force has OTS (Officer Training School).

"The military way," wrote S.L.A. Marshall, "is a long, hard road, and it makes extraordinary requirements of every individual. In war, particularly, it puts stresses upon men such as they have not known elsewhere, and the temptation to 'get out from under' would be irresistible if their spirits had not been tempered to the ordeal."[2] What follows is one account of the "tempering" efforts—endeavors to anneal "candidates" into lieutenants—of Infantry OCS in 1969, during the height of the Vietnam War.

I

In the Army, officer candidates who are not prior service are expected to complete basic and AIT before entering OCS. For me it meant sixteen weeks of training before I even began the Officer Candidate School experience. At Benning, we were "candidates"; that was our rank. Some will tell how difficult Marine OCS or the Platoon Leader's Course is, or how tough boot camp may be; to be sure, it all depends upon the company one is assigned to and the D.I.'s he encounters. I would argue that Infantry OCS, at least during the Vietnam War, was among the most rigorous, stressful, demanding experiences in the U.S. military. Its premier purpose, after all, was to teach young men how to lead an infantry platoon or company in combat. OCS traditionally takes up the slack when West Point and ROTC schools cannot produce enough lieutenants during a national crisis. During wartime, infantry lieutenants find themselves in a growth industry, so OCS normally expands to meet the demand for young ground officers. Although some of the candidates at Benning had initially been drafted into the Army, attendance at OCS was entirely voluntary. To attend, one had to pass a written exam and interview, preferably have a college degree (although only a high school diploma was mandatory) or substantial prior service, and make it through BCT and AIT in good standing. The company I was sent to, 96th OCS Company, was among the most difficult and demanding. Toward the end of our cycle, we won what was described as the "Tiger Tactics" award for the most aggressive and accomplished company from a number of competitors. OCS is an experience I am glad I went through, but I would never knowingly do it again!

Some perspective on my OCS experience ought to be given here. Although I worked very hard in OCS, I graduated at about the middle of my company in academic and leadership achievement. I was never an outstanding candidate in any category, but neither was I, at any time, in danger of being "boarded" (i.e., expelled from OCS as an incompetent). In fact, my one claim to glory during OCS is that, throughout the entire time of my candidacy, I never had to walk a single tour. (A "tour" meant pacing in a given area, rifle on shoulder, for a prescribed time—after passing a rigorous inspection—as penance for various infractions of rules or to work off demerits assigned for unsuitable conduct of one sort or another.) Throughout OCS, candidates rated one another; each week we had to name the top five and the bottom five in our platoon. Candidates consistently in the bottom five on many lists were likely not to be in OCS very long. In twenty-three weeks of OCS, I made the top five of one person's list once. (I did not vote for myself.) I made the bottom list twice (of two people, or twice of one guy[!]). I really was very happy right in the middle of the pack all the way through. Thus, I was no great shakes as a candidate, but neither was my commission ever in much doubt.

Lieutenant "Korea" (in AIT) had warned us about OCS. The only difficulty with his admonitions was that they underestimated, by far, the rigor of OCS. This is not to say that OCS did not have some redeeming virtues: the teaching techniques employed at the Infantry School were excellent, and I enjoyed learning from and about them. The candidates, on the whole, were honest, intelligent (except for the two guys who rated me at the bottom of my platoon in leadership), forthright, energetic, and patriotic young men in whose company I thought it an honor to be. There were a number of funny, even hilarious, moments, some of which I will recount here. But the purpose of Infantry OCS, after all was said and done, was to equip us in mind and in body and in spirit to kill—and that is no laughing matter ever. To find, fix, fight, and finish the enemy—that was our business. "War is our business," was our chant, "and business is good." Graduates of Infantry OCS knew they were headed to Vietnam, where they would have charge of a rifle platoon and be responsible for about forty lives; probably, during our tour there, we would also (given officer attrition rates) have command of a rifle company after some combat seasoning. I never forgot in OCS that, if I failed to learn what I was supposed to, I could be responsible, directly or indirectly, for the death of some young trooper unlucky enough to draw me as his leader. In fact, our classes often started with this adjuration: "You will learn this, or you will die in Vietnam." I accepted that in connection, say, with classes on mortars, but I found it hard to buy in connection with classes on social etiquette.

OCS began for my buddy, Doug Jhoram, and me after our drive to Columbus from Massachusetts. I was as ready psychologically for OCS as I could be; but I was not as ready physically as I should have been (although I certainly thought I was in excellent condition by that time). OCS was intended to produce stress. OCS did not have drill sergeants; rather, it had tactical officers (or tacs), whose job it was to produce battle-worthy leaders. Theoretically, this was a fair idea, but almost all of our tacs were very recent graduates of OCS, with no experience as an officer, let alone as the platoon leader of a combat unit. Frequently the tacs confused legitimate stress (which every human being who has charge of others must learn to deal with) with fraternity pranks and silly harassment. The Army institutionally accepts the idea that it is necessary to educate and train such leaders in a stressful situation so that when combat and the fog of war do come, those leaders will be as prepared as humanly possible. By controlling the calorie intake, the amount of sleep, and the nature of the physical exercise regimen, the tacs could make life extremely difficult for their candidates. Most did exactly that. The tacs, serving a few months at OCS after their own graduation, were just awaiting their orders to RVN—the Republic of Vietnam. A few tacs were excellent; some were good; most were fair; some were sadistic. I knew tacs in each category. And, worse, they knew me.

Soon after we reported for duty, we learned two critical things: everything we did was to be at doubletime. We didn't walk; we ran. The standard position for a candidate was the "front leaning rest position," which I had come to know and love in BCT. We also learned that everything we said had to be preceded by this: "Sir,

Candidate X. . . ." Most of the candidates in the newly formed 96th OCS Company were college graduates, fresh from one AIT or another around the country. A number were "prior service," meaning that they were Army enlisted personnel, up to the rank of E-7 (sergeant first class or platoon sergeant), who were trying to earn a commission as a second lieutenant. We stood at attention in front of our barracks, our first day at Benning, to hear the tacs tell us that we were nothing; that we were worse than nothing; that we were scum; that we didn't deserve to be in the United States Army, let alone at Officer Candidate School; that we would never, ever earn the gold bars of an Army second lieutenant—especially of a U.S. Army *infantry* second lieutenant. Then the tacs concentrated their fire on the NCOs.

I remember the day well, for I was standing behind an E-6 (staff sergeant). At the time, I was an E-2, a private. Most of my drills at Dix had been in the grade of E-6. Now the staff sergeant to my front and I were the same rank: candidates (except that, during OCS, he continued to draw E-6 pay; I was paid at the rank of E-5 during OCS). The tac went over to the E-6 innocuously enough. "Hello, sergeant," said the tac, in a kindly way. "Nice morning, isn't it?"

"Yes, sir," said the hapless NCO. That was his first mistake.

"You blithering dolt!" screamed the tac, only inches from the ear of the NCO. "How many times do we have to tell you that you NEVER open your mouth without the words, 'Sir, Candidate X' emerging? Get down and give me one zero zero (100 push-ups; the tacs would state a number; what the number was connected to became immediately obvious)."

By the end of OCS, even I was able to do about 100 real push-ups—but not this day, and I felt sorry for the straining NCO to my front. I will admit, however, that my dark side emerged here. I enjoyed seeing an NCO "get his." I had never seen anything like this in my military career of five full months—undoubtedly why it was done. We were right back to the pecking order all over again. Who's in charge here? Several months after this episode, I would be an officer in airborne school—only to have NCOs [so-called "black hats," or NCO airborne instructors] politely but firmly dropping me for push-ups! I hate push-ups.

The NCO had done about fifty push-ups before he began to weaken and tremble. The tac screamed, "Recover [get up], you pathetic piece of human excrement. You sicken me! You make me want to throw up! Is that your intention? Are you trying to humiliate me by making me lose my breakfast right here in front of these troops?"

"No, sir," replied the shaken NCO. That was his second mistake.

At these words, two other tacs joined the first one, all three screaming obscenities into the face and both ears of the staggered NCO. They then pulled him front and center, where, in front of the assembled 96th Company, they collectively deplored his very existence. My secret enjoyment of the NCO's plight was long since over. I felt, in fact, a little ashamed of the conduct of these "officers and gentlemen" who could savage a non-commissioned officer in front of troops, even though we were all in this together now. I began to empathize with an NCO as I would never have thought possible.

At this time, NCOs still wore chevrons on their sleeves. This E-6 had been in the United States Army for perhaps ten or so years. He had no doubt earned his stripes, and he was proud of them. Had he been a misfit, he would never have received orders to OCS. Now one of the tacs produced a razor blade. With the NCO standing at attention, the tac slit a thread or two on the chevrons on each sleeve. The two other tacs then grabbed the sergeant stripes and yanked them off the uniform. "You're a piece of dung," announced one of the tacs. "Sound off, scum bag! Are you a piece of dung?"

This created a real problem for this, well, former NCO. If the beleaguered candidate agreed, he could be attacked as spineless. If he disagreed, he could be attacked as stupid. As I recall, the candidate agreed. "Yes, sir!" he sounded off. It was his third mistake. It's not necessary to continue here with the trials of this poor NCO. But I certainly never forgot to begin every remark with, "Sir, Candidate Toner . . . ."

Was this beastly treatment of a sergeant necessary? I suppose that my answer comes as a surprise: yes. Most of the candidates in 96th Company were young college graduates, just out of BCT and AIT, where NCOs were kings. That was no longer true, and we had to be told and shown vividly that OCS was not for or by or about NCOs. The lesson—like many in OCS—was carried too far; and it need not have been done in the way the tacs chose. But somehow it had to be done. OCS was basic training all over again—although a great deal tougher—but this time for officers. At Saint Anselm's I had been a three-year member of an honor society known as the King Edward Society. Its pledges had been put through a period of rather good-natured harassment. Having played sports and having been a member of the Knights of Columbus and other clubs, I was thus accustomed to mild "hazing." It's difficult to survive BCT and Infantry

AIT if you take yourself too seriously! But the kind of personal humiliation and degradation routine at OCS could not, in my judgment, be justified. A few years later in Europe, a friend of mine, also an Infantry OCS graduate, said that, after one week as a candidate (having made only that seven-day investment), he would have killed for his gold bars. The tacs seemed relentless and remorseless in their single-minded quest to get candidates to quit. Those who did quit received assignments in the regular Army; some even earned a promotion (to E-3 or to E-4, depending upon when in the cycle they quit; most had come to OCS in the grade of E-2).

Some candidates who did not quit were nonetheless boarded, as I have already alluded to. When candidates criticized one another in writing—as we had to do—records were kept and counseling sessions were held. If a candidate amassed too many demerits, he was summoned in front of a review board of officers and, in essence, made to show cause why he should be allowed to continue. Failures in so-called "critical" exams, failure to exhibit proper leadership, failure to improve after counseling sessions with the tacs—all these meant boards, which were usually fatal to the candidate. One candidate from Minnesota, Dwight Hughson, was boarded out during his twenty-second week—with ten days to go until our magic graduation day of 8 August 1969. Hughson had already bought his dress greens and dress blues! Tacs had to justify with paperwork those they condemned. It was, therefore, much easier on the tacs if candidates they wanted out simply quit; there was little or no paperwork then. Occasionally, a candidate from a company ahead of us in the cycle, say in the twelfth week or so, would have to be re-cycled, or sent back to another company, because he missed some training due to an injury or family problem. It was well known that 96th Company did not take such rejects or re-cycles from other companies. Such candidates were harassed with deadly effect—almost literally.

One re-cycle, Candidate Golman, became a tac target after he entered the 96th Company. One evening, after extensive verbal abuse intended to inveigle Golman into quitting had failed, Golman was taken to the latrine, where the tacs had turned on all the showers up to full blast on hot. Golman was put inside a mountain sleeping bag and zipped up, with only his head showing. The latrine was by now a cloud of hot steam, and all other candidates were kept out. But we could hear the screaming from the latrine all throughout our platoon's area. "Knock 'em out, Golman," screamed a tac, "two zero zero!" Golman must have been trying to comply. "You gonna quit, Golman?"

The screams and taunts could be heard all around the floor. None of us interceded. None of us tried to help Golman. Golman's tormentors were commissioned officers of the United States Army, whose professional job it was to look out for the welfare of their men. We, as candidates of the 96th Company, were their men. Golman was their man, their responsibility. Golman passed out. A tac threw water in his face. "Get up, Golman, you pussy! Knock 'em out! You gonna quit, Golman? Quit, you bastard, or you'll knock 'em out until you die!" Golman passed out again. More cold water in the face. More insults. At last, the tacs departed, leaving Golman semi-conscious on the latrine floor. He never quit. On a subsequent "Ranger Problem," as it was called (or FTX, for Field Training Exercise), Golman kept passing out. Apparently, he could not sweat enough to cool his body. I do not know the medical reason. But I know that Golman graduated from OCS and received an assignment to Okinawa. What happened to him after that, I have no idea. Nothing more was ever said or done about his torture that night in Infantry OCS.

All these fine young men whom I have alluded to knew what Golman was going through. But not one of us did anything to intercede. I guess we figured it was Golman's problem. In fact, we were moral cowards. I should have stormed into the latrine—somebody should have—demanding to know how U.S. Army lieutenants could act that way. What they were doing was criminal; they could have killed Golman. It is true, by way of mitigation, that we had been ordered out of the latrine, told to stay away. But we knew what was happening. We supinely and spinelessly acquiesced in accepting those orders, illegal and unethical as they were. William Calley, Jr., who perpetrated the massacre at My Lai during the Vietnam War, was a graduate of OCS. What happened in that latrine was hardly on a scale with My Lai. But the parallel, to the extent that any exists at all, is deeply disturbing to me, for I said and did nothing that night to prevent or to stop what was happening. That young man's only offense was that he wanted to be an officer in his country's Army as badly as those "officers" wanted to prevent him from achieving his goal. After the incident, I reflected morosely about my failure, and I resolved never again to fail to speak out when I should. I had been taught another positive lesson in a viciously negative way.

II

A number of movies have told stories about training brutalities. I do not know whether some of them were based in fact or made up for entertainment. Golman's case was real. There is, however, another side to the story. No one in my OCS class ever knew that I did this but, because I was deeply bothered by Golman's torture—and by the fact that I had gutlessly permitted it to go on without so much as a word of protest to the tacs—I resolved, in the comfort of time well after the event, at least to find out why it had occurred. A few days after the episode, I went to see his chief torturer, Second Lieutenant Forrest. I knocked on his office door and asked to see him: "Sir, Candidate Toner. Request permission to speak with you, sir."

"What do you want, Toner? Make it fast. I'm busy. Assume the position."

"Sir, Candidate Toner. Yes, sir!" I dived into the front leaning rest position in front of Forrest's desk. Our conversation, as best I can recall, went something very much like this:

"Sir, Candidate Toner. Was it really necessary to treat Candidate Golman the way you did the other night, sir?"

"Do you really think that is any of your business, candidate?" Forrest replied.

"Sir, Candidate Toner. I just wanted to ask you about it, sir. It didn't seem right, sir." If the constant vocative *sir* seems rather sycophantic, we were taught that every sentence, even after our identification line, would begin or end with "sir." As I recall, however, we were given a choice about whether to begin or to end the sentence with that form of address. In the conversation, I was not trying to be obsequious but rather following protocol.

"Get up, Toner," Forrest ordered. "Stand at ease."

"Sir, Candidate Toner. Yes, sir. Thank you, sir." I had no idea what to expect, for this was a departure from the routine. Usually counseling was done with the candidate in a push-up position. It was a break if we were not required to knock out push-ups while being "counseled." I did not know what to expect.

"Knock off the bullshit, Toner, and just stand there. You asked me a question, so I'll answer it. I don't think I owe you squat, so don't feel as though I think I have to answer you, let alone explain anything to you. I'm going to tell you something that nobody else here knows. You listen; then get out of my office and shut your mouth about this. You understand me?"

For my first and only time in OCS, taking advantage of Forrest's offer, I responded as a soldier rather than as an Officer Candidate: "Yes, sir."

"Toner," Forrest began, "my brother enlisted in the Army. Infantry. He died in Vietnam because a lieutenant, ROTC type, couldn't read a fucking map. That lieutenant killed my brother because he was an incompetent asshole. I don't want this company turning out incompetents; you understand me? I don't want them killing other people's brothers. I don't want them killing me. Now get out of my office."

That was it. No details. No discussion. No necessary connection to Golman. I am not a psychiatrist, and I don't pretend to understand the full meaning of that revelation. Forrest was about twenty or twenty-one years old. His father was an Army colonel, then stationed in Hawaii. Forrest's brother had died in combat in Vietnam as an Army infantryman because of the alleged negligence—I know absolutely nothing more about the case—of a lieutenant. Forrest therefore had his reasons for his sadistic treatment of candidates—there were plenty of other examples besides Golman. In my view, he was too young (he himself had enlisted, but he passed the OCS test) and too immature to train officers. But the fact is, I guess, that his motives were good. But as de Gaulle once said to Petain after the latter's collaboration with the Nazis in German-occupied France during World War II (Petain having contended that his intentions, at least, were good): "Only God can judge intentions; men must judge by consequences." In a way, Forrest reminded me, despite his noble sentiment, of the E-2 who had made us dry shave at Fort Dix. Power was too much for him. It is too much for a number of officers.

Forrest and one of his buddies also did something that I thought was disgusting. They would look in *Army Times* every time it came out, getting the names of all the soldiers from the Columbus, Georgia area killed in Vietnam. They would then call the widows of these soldiers to offer, well, "consolation." A couple of years later, Doug Jhoram was visiting me in Germany. He had read in that same newspaper, the *Army Times*, that Lieutenant Forrest had been killed in Vietnam. Jhoram laughed about it, thinking that he had probably been fragged (killed by a fragmentation grenade) by his own troops. I regretted it but could not bring myself to express great sorrow. Forrest was a kid, and he had been asked to do far more than his maturity permitted.

That may have been true, too, of OCS's best-known product, Lieutenant William Calley, Jr. Calley and I went through OCS at different times, but the training was essentially the same for both of us. The training at the "Benning School for Boys" (as it was called) could be degrading and, as I have outlined, it could teach the wrong responses to authority (and to some extent it could have contributed to the wanton slaughter at My Lai perpetrated by Calley and some members of his platoon).[3] But nobody ever told me to take unarmed and unresisting men and women and children, put them into a ditch, and murder them. Calley and his men deeply stained the uniform of the United States Army. Although I know that some might object that it is very easy for me, a non-combat, Vietnam-era veteran, to write those words, I am also an American. When Calley is held guiltless for his crimes, we are all held collectively guilty. If some of the stupidity of Infantry OCS training contributed to Calley's vile deed, some of the training at Benning—and ordinary decency and fundamental right reason—should have prevented that massacre. Whether I would have had the sense and the courage not to commit such a wanton crime—even if directly ordered to do so by superior authority—I will never know. But I hope so. I believe so. I pray so. The problem of producing misfit lieutenants is not solved by insisting upon their having college degrees, for the degree, in itself, means very little. Perhaps, however, the graduate is a little older, a little more mature. The non-college OCS graduates I knew tended to be somewhat less stable than the OCS college graduates. Certainly, from my vantage point now, I believe that lieutenants should hold degrees—as well as endure the stress and strains of OCS-type training (intelligently managed by reasonable and experienced officers, without Lieutenant Forrests around to harass and humiliate them). Anyone who spends a few years in the Army will meet OCS officers of various quality. But the same is true of ROTC officers and of West Pointers. What matters is the quality of the man or woman, and not the source of commission—a problem discussed later in this book.

III

My own tac officer, Lieutenant Dan Joseph, had the unnerving habit of telling us that "You'll come home from Vietnam in a bodybag." On balance, though, he was a reasonable tac, and I ended up going through airborne school with him. I have no idea what became of him after airborne school. But I recall with a chuckle one minor occurrence in OCS involving Lieutenant Joseph. One day, at retreat, in about the nineteenth week of OCS, the student company commander (taking his turn in that candidate leadership position) was absent from the formation. Lieutenant Joseph asked the candidate executive officer (the second in command) where the student company commander was.

"Sir, Candidate Smith. Candidate Jones is turning in a compass, sir!"

Without missing a beat, Lieutenant Joseph responded, "I know Jones is thoroughly incompetent, but I want to know where he is!" Several of us started to break up, which was unheard of at OCS; and it was only with the greatest of effort that I restrained myself, as did the others.

One of the other tacs, generally regarded as excellent, left OCS for Vietnam in time to participate in what we now know as the battle for "Hamburger Hill." We received word, not too much later, that he died there. We had occasional laughs, it was true; but there was much greater opportunity for tears.

Daily life in OCS was incredibly demanding. We slept little, and we were allowed to exchange sheets only after more than *three months*. They did get somewhat nasty in that time! We rarely slept more than six hours a night and sometimes less. But worse than the lack of sleep was the regimen in the chow hall. We had to eat at attention, meaning that we looked straight ahead, never down. Eating in such a manner can be very frustrating. Try it. After every bite, we had to put our forks down and chew. Of course, conversation was forbidden. We also had a limited time to eat. To ensure that we had enough "fuel" to get to our next class and through our day, the tacs would announce, with about three minutes left in our scheduled time for eating, that we now had "pig privileges," meaning that we were to put our faces right over (or in) our chow, consuming as much as we possibly could in the two or three minutes remaining. This drill often transpired as we walked up to the tray-return area where we had to separate utensils, garbage, and paper goods. I never saw a single fat candidate—at least after three or so weeks. Perhaps the only good thing about this was that we never had KP. There were civilian contractors to do that chore; it was thought that OCS students should not have to miss classes to do dishes. After my time at the Reception Center and at BCT and AIT, I was glad of that, at least.

We dreamed of food. Occasionally we tried to get it by unauthorized means. Unauthorized food was called "pogey bait." One night, a few guys took up a collection and made a run to the PX, buying candy bars. They used the best tactical means to ensure the success of their mission: reconnaissance and patrolling, scouts and intelligence, communications and security. They even used the five-paragraph field order familiar to all infantrymen. Still, the tacs caught the intrepid raiders on their return from the PX. Von Moltke is supposed to have said that no war plan survives first contact with the enemy; ours didn't. There were candy bars for the entire company—two per candidate! I do not recall how many guys were left, by this time, in the 96th Company—say 100. That was 200 candy bars. The patrol leader was identified. The tacs said that, because he was hungry, he would be satisfied. He could eat the candy bars while the whole company watched. Imagine eating 200 candy bars! Superman—with super belly—couldn't do it. Still, this candidate had little choice, for he had received marching (munching?) orders from a commissioned officer. Gamely, he plunged ahead, eating chocolate, peanuts, marshmallow, caramel, and many other gooey substances. Until his stomach rebelled. He threw up. He did not throw up a little. He threw up a lot. The tacs, undaunted by this evasive tactic on the candidate's part, insisted that they would not be put off. "Candidate, that ruse isn't going to work! You *will* eat those candy bars! You *will* re-eat what you have 'discarded.'"

I remember thinking, "Oh, no. They can't be serious." They were. The poor "patrol leader" began to eat the candy again, beginning with his own vomit. As I watched, standing in formation, my own stomach began to experience distress, and I am sure that the stomachs of my fellow candidates were similarly bothered. The patrol leader barfed again.

"No good, candidate, no good. Are you refusing a direct order, candidate? Do you refuse to eat the *unauthorized* pogey bait you have here, candidate? And all we're trying to do is get you fed! We want you to have the fruits of your labor, candidate. And you refuse! Eat it! Suck it up! Do it! Now!"

After a while, the tacs tired of this disgusting enterprise, so they brought the by-now reeking offender into the barracks. They took the remaining candy bars (plenty were left) and went to the hapless patrol leader's platoon area. They then took a buffer and, after mashing up the candy bars, buffed them into the floor, smeared the mess on the walls, and otherwise distributed the chocolated goo all over the platoon area. They then said good night, after telling us that there would be an inspection at 0530 hours. Of course we spent the night trying to get that mess out of our floors and off our walls. It took me a very long time before I ever wanted to eat another candy bar.

We went through OCS in our platoon area, thirty guys or so, using one urinal and three commodes. We had more, but we never used them, for there was less to clean that way. Everything had to be immaculate all the time. In OCS there were usually three persons in a room. I was the only person I knew who spent six months with the same roommate, Larry Leland from Arizona. Larry and I had occasional other roommates, but we went through OCS together. Larry was married, so after about twelve weeks of OCS, Larry got to visit his wife. Candidates had been told that it was unwise to have their wives at OCS. My fiancee was still in college in Massachusetts, so I had no problem. In fact, I was as close to my fiancee as Larry was to his wife, and she was living in Columbus for most of OCS.

Larry and I, and most of our friends, developed a system according to which we stepped on only a few established tile squares in our room. That way, we had to buff only those few tiles every morning before we left our room. Of course, our boots had to be spit shined; our fatigues had to be pressed (I owned about fifteen sets) all the time; and our brass had to gleam. The OCS haircut was a landing-strip cut, meaning that there was one swath of hair cut from front to back (thus the "landing strip") and the sides were very short.

We did learn a few tricks. Most of us kept two sets of items, one for use and one for display (for inspection purposes). The honor code at OCS prohibited lying, but it did not prohibit getting around things. For example, we would use our display toothbrush and display toothpaste once. Then when asked, "Have you used these?" during an inspection, we could reply, "Sir, Candidate Smith. Yes, sir!" We learned to sleep whenever and wherever possible. We got pretty adept at eating at attention. Wives got pretty good at surreptitiously meeting their husbands for, literally, a second or two to exchange laundry. I can hardly sing a note, but I sang in Catholic men's choir every Sunday so that I could go to practice on Saturday whenever training permitted. The chaplain was great. He always had lots of pogey bait for us. The first time he had candy bars, however, I just couldn't handle it. With great effort and persistence, though, I managed, after a while, to get down a Snickers or two at every choir practice.

A kind of pride did attach itself to our efforts. We marched to classes in OCS and, after a while, our marching was accompanied by songs and/or a snare drum. We marched "standin' tall." When an Infantry OCS company marched, it marched well. We carried clipboards (with "Win in Vietnam" stickers on them) for note-taking, but our left-hand holding the clipboard never moved. Our right hands were synchronized so that, from the side, the spectator saw only one right hand swing up together. All of us got proficient at "22-5," the Army manual explaining drill and ceremonies. We learned to blouse our boots (trousers, really) in the OCS way. Enlisted personnel used blousing rubbers to hold their trousers at the top of their boots; we tucked them down into the boot (most of us had bought the Corcoran "jump boots" which are a little higher than G.I. regular boots), pulling the trouser leg closely together at the back so that it looked sharp.

Inspections were continuous and meticulous. Although I escaped walking any tours while I was in OCS, I attribute that more to luck than skill on my part. If a tac wanted to give someone demerits, he could almost always find a way. I also escaped from serving in the worst imaginable position in OCS: candidate company commander. The C.O. was under constant scrutiny, and one mistake led him to the tacs' noose. Of course, occasionally the real company commander, a senior captain and combat veteran, would inspect us. During one inspection, the captain was examining our weapons. He arrived in front of my buddy, Doug Jhoram, and began inspecting his rifle. "What's this, Jhoram?" I heard the captain ask.

"Sir, Candidate Jhoram. Duuuuuuust, sir!"

"That's not dust, Jhoram. That's fiiiiiiiilth!" replied the C.O. Jhoram spent many subsequent hours with his weapon, cleaning it to the standards of the divine. There was a point to ensuring that candidates, future infantry leaders, knew their weapons (and cleaned them) religiously. We knew that upon our training depended our lives and those of the "snuffies" we would lead. No one objected to tough, realistic training. We wanted to survive in combat.

IV

Early on in OCS, we candidates underwent an interesting and, I think, very useful training exercise. We were brought out to a field or range area known as the leadership reaction course. The LRC is a large collection of what might be called puzzles. The company was broken down into small groups; within each group a different leader was appointed as the group moved from puzzle to puzzle.

The leader read the nature of the problem, and he was given the necessary equipment to deal with it—if he could devise a course of action. A time limit was placed on each problem. We knew that all the problems had school solutions; but the time was short and the pressure was high as the guys in the group looked squarely to the leader to resolve the problem quickly and efficiently. We had to think on our feet!

I can recall now only the problem I had to solve when I was the leader. We had to climb up a ladder onto a platform overlooking a small pond. Across the pond was another platform. A sturdy cable was stretched between the two platforms. My job: to get my group across to the other platform. The problem was intensified by instructions requiring me to have one of my men play the role of a severely wounded soldier, unable to move under his own power. Various items of equipment, such as rifles, were on the platform; everything was to be transported across to my objective. Nothing was to be left on my platform when time was called.

As I read the problem, I thought, "I haven't got a clue. I do not know how to solve this thing. Now what do I do?" First, I was honest. While trying hard not to give the appearance of confusion, I called upon my buddies for ideas, suggestions, recommendations. It was our problem, not just mine, although as the leader I would have to make the final decisions. We got ideas out on the table. I acted as a kind of seminar director, trying to sort quickly through ideas, picking elements of the best, discarding the worst, weaving them eclectically into a practicable plan. Time was very short! We sent over the small equipment by tying it into (or even with) our fatigue shirts. (You always have something with you that you can use: belt, socks, shirt.) Most of the group then went hand over hand across the cable to the other platform. That was simple enough. But we had to get our wounded man across.

He couldn't pull himself. One of the items on the platform was a stretcher—of sorts. As squad leader, I went last, checking out the platform to make sure everything had been taken. (That seemed natural, but it was one of the lane grader's checkpoints on this leadership problem.) One of the guys in the squad was very big and strong, and I had picked our smallest guy to be wounded. I used the big guy as an ox, tying the stretcher to him. He went across the cable using his hands and crossed feet, pulling the stretcher behind him. I followed it, giving it an occasional shove or push when necessary. A couple of larger items that the guys couldn't carry themselves were lashed to the stretcher. Just as time was running out, we reached the other side.

We did a check on the other side "for the record" to be sure we had everyone and everything we were supposed to have. That was the last element on the school solution checklist. I got credit, therefore, for solving a problem that had initially completely puzzled me. Because few other problems on the LRC were solved that day, I looked pretty good. There was even a P.A. system there announcing which groups, under which leaders, solved LRC problems, so I got some early good publicity—not a bad thing when you're rated and re-rated constantly by your peers! But, much more to the point, I learned some very positive things that day about leadership. And, this time, I had learned them from a positive, practical leadership exercise.

First, I learned that there is something to be said in favor of the old Infantry School adage, "Don't just stand there; do something!" Leaders must be seen as leading; they must create a proper image. I do not mean a phony or a false image. But appearance, as the Army has taught for years, does matter. Action does matter. Direction and vigor do matter. One hopes, of course, that these are not just vacuous, meaningless activities—false facades, as it were. But a leader who will not lead will have followers who will not follow. Had I displayed externally the confusion and indecision I was feeling internally, I think that my followers would not have responded as enthusiastically as they did to my invitation for a pooling of ideas. Moreover, they would have regarded me as inept and incompetent. *I* solved nothing; *we* solved the problem. All I did was to make the necessity for consultation a virtue. This lesson has been invaluable to me as a teacher and a coach.

Second, I learned that a leader must be prepared to listen carefully after taking the action and issuing the orders inviting his followers to participate in planning.[4] During the French Revolution, someone is supposed to have said, upon seeing a large crowd coursing down the street, "I am their leader! I must follow them!" In part, leadership consists of developing people's talents and then helping them to channel those enhanced talents in the proper direction and for the right purposes. One can jump on a horse and ride off in a number of directions, but the key element of sound leadership is knowing where to go before mounting up![5] Sometimes, in that connection, the wise leader is prepared to listen attentively to his subordinates. That need not mean, in the least, that he become servile to their judgment or to their opinion. But he must know how far he can go, how much latitude he has, and how much rope his charges will give him. (Even the President of the United States can lead only so far; he must be sensitive to public opinion while simultaneously trying to influence it.) In listening to my peers that day at the LRC, I actually allowed them to solve a problem that, by myself, I had neither the intelligence nor the strength to solve. All I did was to combine into a reasonably coherent window of opportunity the shards of ideas they presented me. Not too long after I went through the LRC, I applied some of its lessons on a German autobahn after a potentially catastrophic nuclear accident involving a convoy of which I was in charge (about which more later).

The Prussians, as I understand it, would commission no officer unless and until he could win the respect of his peers on courses similar to the LRC (and on the "tactical walk," a stroll along terrain during which the would-be officer pointed to certain terrain features as evidence that he could apply in the field many of the lessons he was learning in his classes). OCS employed a roughly similar approach in its map reading and tactics courses.

To me, leadership is *the ability to inspire appropriate conduct beyond the expectable*. That's it. One can inspire in a number of ways. But the key element consists in knowing what is appropriate—which requires vision—and in knowing, first, what will ordinarily be done (requiring no special leadership of the group); second, what extraordinarily must be done by the group in this situation (requiring special leadership); and, third, how to respond to the challenges of the "leadership margin" between the two.

Thus I came that day to an image of what leadership is all about. The effective leader adds to what his charges already have. There is highfalutin talk about transformational and transactional leadership,[6] and it's worth reading about in academic circles. In the field, however, what matters is what the leader can add to whatever his charges already have, or how he can orchestrate their knowledge and experience. Leadership is always done *at the margin*. If troops will do the ordinary, the leader, at times, must inveigle them into doing the extraordinary. But the leader must know the critical difference between that and the impossible, which, by definition, not only cannot be done (the Seabees' motto notwithstanding) but should not (except in the one-in-a-million case) even be tried, for it will result in chaos, or in resentment, or perhaps even in mutiny.

For example, no high school track coach, now matter how good, can get a track star to jump over the moon. To attempt this is simply asinine. The leader, therefore, must set goals. (If he fails to do that, he has failed in his fundamental leadership challenge.)[7] But they must be sensible and prudent and achievable goals. To tell one squad of soldiers in 1942 that they alone would seize Berlin would have been foolish. To tell a high school pitcher that he must throw his fastball at 200 miles per hour is beyond the realm of credibility. But for a coach to preach the need to that pitcher for location and movement and velocity in his arsenal—there is the potential for leadership.

If the leader expects only the routine, he has eliminated the need for himself. If he expects the miraculous, he has created the need for God. The trick is to develop the tactics necessary for getting out of one's subordinates (whether soldiers or students or musicians or athletes) more than they routinely think they can do but not to the point that reason and science counsel is super-human and that will break the spirit of one's subordinates.

Leadership can be thought of, in economic terms, as "value added." There are two common-sense expressions governing this. The leader cannot turn chicken shit into chicken salad. But he can make lemonade out of a lemon. Knowing the difference between the two is vital! The good leader, I learned in OCS, can get people to give blood; he cannot get them to give all their blood! Cadets and candidates may well expect miracles from God; but they have no right to ask miracles from their men. Of men, the supernatural should not be expected—but the excellent (that is, that which excels; that which transcends the normal, the average, the mundane) should be insisted upon! In a phrase, the good leader must strive to make the excellent routine! In the seventh grade I had a math teacher who told me that "Habit is a cable and we weave a thread of it every day." Aristotle said much the same in his writings—that we are what we practice. Baseball and football coaches have said for years that we play as we practice. As one author said in a *Parameters* article, the heart of leadership is the four C's: competence, candor, compassion, and commitment.[8]

V

After twelve or thirteen weeks, we turned black, meaning that we became intermediate candidates. At about twenty weeks we turned blue, meaning that we were Senior Candidates, entitled to be saluted and called "sir" by lesser candidates from another company. Turning black and blue (how appropriate, given all the punishment we went through!) are highlights of my life. I don't mean that they compare with other, far more personal, events, but I will never forget them. My fiancee flew down to Columbus for my Turning Blue Ball, during which time we were able to leave post for a while. We had the formal dinner and the ball; we had a pass-in-review parade; and we were even able to relax some. In the Army, however, any PDA was strictly out! Public Display of Affection meant any hand-holding, hugging, or—perish the thought!—kissing. Army wives were being trained, too. To this day, I have trouble holding an umbrella, which Army officers are not supposed to do; nor are Army officers supposed to hold babies—unless their wives are in desperate need of help. Unmanly, you know.

We still had our last FTX, called a Ranger Problem, to get through. We had spent plenty of time in the field, so that was hardly new. But this was a combat week, with lots of patrolling and no sleep and very limited food and water. For seven days I was to wear the same set of fatigues. When I came back, the fatigues stank so much, I threw them away. Most, if not all, infantry officers now go to the eight-week Ranger course at Fort Benning. Our training in OCS for twenty-three weeks included a tremendous amount of patrolling and squad/platoon/company tactics, obviating, in many people's minds, the need for more of the same in the Ranger course. (OCS graduates did not go to the Infantry Officer Basic course because they had had it; did they ever!) Of course, some OCS graduates did go to Ranger School, but most had had sufficient training at the time so that the Ranger course, for them, was redundant. Now, OCS is only fourteen weeks long, underscoring the wisdom of requiring Ranger School even for OCS grads. So the Ranger Problem held no fear for me because I had already had about twenty or more weeks of lack of food and sleep. But I didn't like the lack of water. We were issued quinine tablets, and we took our water from scummy pools—the only scum I encountered at OCS—in the swamps at Fort Benning. The hot, steaming water actually tasted good—if that was all you had to drink in the July heat at Fort Benning.

One of my OCS classmates, Bucky Pierce, from Brown University, was a sharp candidate, but exhaustion overcame him during this extended problem. We had to dig in every time we stopped. After a few sleepless nights and the heat of a Georgia summer, Bucky was digging a foxhole. He began to whimper in a tiny, high-pitched voice, "Help me! I can't get out!" We ignored it, thinking it was a feeble attempt at humor, especially since he was dug in only up to about his knees in height. "Help me! I can't get out!" he kept insisting. We realized, after a while, that he was serious: he thought he couldn't get out. He had had a stress breakdown, and he was evacuated by helicopter to Martin Army Hospital for rest and for psychiatric evaluation. Bucky was O.K., just spent; and he graduated with us.

On patrol one night, I was the compass man, meaning that I walked second behind the lead or point man, helping to direct him. Light discipline (no lights) was, of course, in practice. Point men in that area were instructed to use a walking stick to test the ground, while still keeping their rifles ready. My point man, apparently, hadn't taken the advice. It was very dark at about 0200. I was watching the two strips of white tape on the back of his soft cap (I could hardly see him) when, all of a sudden, those two strips fell straight down toward the ground. I froze. There was about a two-second delay, and my point man let out a scream as he tumbled into an unseen ravine. As I realized what had happened to him—and to those two white strips of tape on the back of his soft hat—I was seized, in my exhaustion, with a laughing fit. I had enough sense to try to control it and to worry about my buddy, whose arm had been broken in the fall. But remembering "the case of the disappearing strips" still brings a smile to my lips even today. He, too, recovered and graduated with us.

We had all passed the so-called critical exams in leadership, in map reading and land navigation, in mortars, in small arms, and in infantry tactics. A number of other exams and tests were given, too, most of them "practical" whenever possible. OCS emphasized those practical skills: terrain walks, leadership reaction courses, extensive land navigation problems, and the like complemented our constant rotation into various leadership positions (squad, platoon, or company) so that the unable could be weeded out. But on 8 August 1969, all that ended.

That was the day I walked across the stage at Fort Benning as a Senior Candidate. By the time I reached the other side, I was a Second Lieutenant of Infantry, commission in hand, along with a diploma from the U.S. Army Infantry School. My commission was in the U.S. Army Reserve, for only a handful of ROTC graduates and all West Pointers get commissions in the Regular U.S. Army. Of the men I graduated with that day, only a few have re-appeared in my life. Some stayed in the Army and are still in active service today. Some did their active duty and stayed in the reserves. Most (like me) left the Army after their tour of active duty was over. All of us had shared an incredible journey, one unknown to our brother ROTC or West Point officers. Because so few of us OCS graduates remain on active duty and retire from the Army, the OCS experience seems less important to us as the years wear on. We forget and forgive.

But there is something about the depth of that experience that its graduates can share when they meet. The OCS graduates of the

Vietnam era underwent the longest, continuous, straight military combat preparation for officers in the history of our Republic. The tragedy is that OCS is identified with Rusty Calley instead of with a faceless soldier, black or white, who volunteered to accept the responsibilities of leadership as a commissioned officer, doing his level best to serve our country by accomplishing its missions and looking out for its men. For me, the graduates of OCS have the faces of classmates, some of whom died for our country. I am comforted by the knowledge that, even today, other OCS graduates still live for it.

The Fort Benning "alma mater" was meant to be sung rather tongue-in-cheek:

> Far across the Chattahoochee, to the Upatoi
> stands our loyal alma mater, Benning School for Boys
> onward ever, backward never, faithfully we strive
> to the ports of embarkation: Follow me, with pride!

Many took that adjuration about "Follow[ing] me" seriously indeed (even though, at 96th Company, we once had a shoulder patch made up that said, "Quit Pushing!"). Few received thanks for their service in Vietnam or elsewhere, or recognition of what they went through to wear those coveted gold bars. But, in the great majority, they bore their responsibilities well. Robert E. Lee once remarked that "Duty is the most beautiful word in the English language."[9] OCS graduates fulfilled the call of duty. That may be, for them, enough.

VI

My OCS company was right across the street from the U.S. Army Airborne School. In 1969, airborne training was being conducted at Fort Bragg, North Carolina, as well as at Fort Benning. But the chief training center for airborne troops, paratroopers, or sky soldiers has always been at the Infantry School. For six months, I witnessed, from a distance, the training of paratroopers. I couldn't help thinking of the night, now seemingly so long ago, when I was at Saint Anselm's and was involved in the dueling songs—Staff Sergeant Barry Sadler's "Ballad of the Green Berets" against Donovan's protest song, "The Universal Soldier." Sergeant Sadler's song called upon us to "put silver wings on my son's chest [to] make him one of America's best." I thought that my gold bars would be nicely complemented by silver wings. I knew I would receive more respect from the troops if I became a paratrooper.

It didn't look, however, as if I could get a slot in airborne school unless I went "indef." That was the Army term of "voluntary indefinite." After OCS, I had a two-year active duty commitment. If I went indef, I would have an open-ended contract with the Army. My service would end only when the Army wanted it to. In return, however, I could have my choice of post-OCS schools and a duty assignment of my choice (within certain limits) for at least a year. I had always wanted to see Europe and, because I was going to be married after OCS, I thought starting our marriage in Europe would be a good thing. And I could go to airborne school.

My buddy, Doug Jhoram, took the same option I did, but he added Ranger School to his list. After nearly a year of training, I was not interested in another two months at Fort Benning beyond airborne school. So off we went to Airborne School, which is just three weeks. Actually, it was a little easier for us because now we were commissioned officers, living off post in a trailer. We came in early and went home in the evening. The enlisted personnel stayed in the barracks, where they were available for details or jobs.

Airborne soldiers were used successfully by both allied and axis forces during World War II. The Germans were the first to employ the sky soldiers during their attack upon the Netherlands, and they used paratroopers again in their 1941 seizure of Crete. Corregidor was re-captured from the Japanese by paratroopers, and airborne assaults were similarly mounted on the Netherlands, on Normandy, and on Sicily. A special American airborne spirit was born during those dark days,[10] and it is conveyed to airborne soldiers today.

Although most of the people trained at the Airborne School are U.S. Army soldiers, some foreigners are trained there, too, as are volunteer sailors, Marines, Coast Guardsmen, and Air Force personnel. All Special Forces personnel must be airborne-qualified soldiers. Many U.S. Army Rangers are airborne qualified, but it is not mandatory for Rangers. (Some are "straight legs," as the expression goes, or non airborne-qualified personnel.) Officers in airborne training are treated virtually as enlisted personnel, except that after a black-hat airborne instructor tells the officer to do push-ups, he follows it with a respectful "sir."

The Army has pretty much perfected its airborne training, a major portion of which is actually intended to instill a "can-do" spirit. When

I went through airborne school, I was expected to return every salute I received ("Airborne, sir!") with "All the way!" The word *airborne* attaches itself to practically everything one does during the three-week course. The idea, of course, is simple. Airborne soldiers, Rangers, and Marines are supposed to be the elite of the military—Green Berets and Navy Seals have somewhat different, and more specialized, jobs. Some of the "macho" things, appearing rather juvenile at times, are intended precisely to foster the kind of camaraderie necessary to maintain unit cohesion and morale among those forces, which can face the toughest challenges of battle. Therefore, distinctive badges or boots or covers (hats) and the like are used by these fighting forces. In winning the coveted Ranger tab, would-be Rangers have about two months to imbibe the Ranger spirit that "Rangers lead the way." Airborne soldiers must learn that spirit in only three weeks, so the "Airborne!" yell, a maroon beret, and the spirit of the airborne suffuse almost all activities at the school.

The first week of Airborne School is referred to as "Ground Week," for it emphasizes challenging physical training. A great deal of time is spent making sure that would-be paratroopers are in excellent physical condition. There is not time to get them into shape; they are supposed to arrive at the school in top shape. The runs are long, but are done at a fairly slow pace. After OCS, I found them to be easy. The pull-ups can be demanding. (For example, to graduate from Marine corps boot camp today, the boot must do two pull-ups. In airborne school, the black hats usually insist upon eight good pull-ups, but that varies by company.) The calisthenics are similarly demanding, and airborne trainees get a great deal of sawdust in their fatigues (now known as BDUs or as Battle Dress Uniforms) from extensive work-outs. But at the end of the week, trainees are physically ready for further training.

When I was in Airborne School in August 1969, the class had a high number of lieutenants because the West Point Class of 1969, coming off its post-graduation leave, was beginning its airborne and ranger training. There were enough of us officers so that we collectively decided that no one would fall out of a run. Some of us lieutenants therefore brought up the rear on every run. If anyone began to fall out, we would grab the poor trainee on each side and persuade him that he really didn't want to quit the run. If need be—and occasionally it was—we escorted him on the run. That was the kind of unit spirit that airborne training encourages. (I understand, however, that preventing troops from falling out of runs in airborne training today is no longer permitted, lest the would-be dropout be hurt by being made to continue running. But if someone drops out of two runs, he is re-cycled or dropped from the school.) In Airborne School, the attrition runs at about twenty percent—varying from company to company and cycle to cycle.

The second week of Airborne School is referred to as "Tower Week." After extensive instruction on a number of ingenious training devices, trainees are brought up into a 34 foot tower to practice their exit from the aircraft. Now, 34 feet doesn't look too tall—*from the ground*. But I assure you that it looks plenty high when you're about to jump out the door, depending completely upon the harness you have strapped to your body and connected to a cable. Every now and then somebody freezes in the door and refuses to jump out of the tower. Such refusals normally result in some counseling by the instructors. If the student refuses again to jump, he will do some details for some days or weeks until he is sent to a straight-leg unit.

Some of the enlisted personnel I got to know in Airborne School had dropped out of high school. Their completion of basic, AIT, and now airborne school was the highlight of their young lives. They were all volunteers for airborne training, and normally they were a dependable and mentally (and physically) tough crew; despite their earlier scholastic problems, they were not quitters. Indeed, the Army had given them a sense of accomplishment and pride. They tended to encourage one another very well without added comment from the student airborne officers. One of the student airborne officers was Lieutenant Joseph, my OCS tac, now on orders to Vietnam. Now we were equals. I tended, however, to stay pretty much away from him, talking at breaks with my OCS classmates or with the West Pointers.

After completing the 34 foot tower, we eventually moved on to the 250 foot tower. The would-be paratrooper hitches up to the parachute, which is connected to a large cone-like object. He is hauled up 250 feet off the ground, before the chute disengages and he floats to earth. I am told that the view from the top of the tower is breathtaking. I really can't recall. All my breath was gone by the time I got to the top, and I wasn't interested in the view.

During these two weeks of training, speakers at the school incessantly blared out the Paratroopers' Song, "He Ain't Gonna Jump No More." The song tells the story of one paratrooper whose parachute fails, resulting in his unfortunate, rapid, and total collision with the ground. The lyrics, sung to the tune of the "Battle Hymn of the Republic," suggest that "intestines were a-dangling from that paratrooper's boots" and "Gory, gory, what a helluva way to die"—so that he ain't gonna jump no more! The song deals in a humorous way with the actual danger of death. The airborne school is serious business, but the song removes some of the sting and worry that every paratrooper feels. The Army trains thousands of parachutists every year, and injuries are rare because of the great care the instructors take in the training of every airborne candidate. But the song is known well to all of us!

By week three—"Jump Week"—the soldier is ready, if not entirely willing, to put his training to the test. To earn the coveted silver wings, he must complete five jumps from two different types of aircraft. Today's paratroopers actually walk (take a step) out of the door of one aircraft from which they are said to "jump." Little attention, however, is focused on the aircraft. As the jump masters remark, the jump itself never killed anybody; it's the landing that does that. PLFs—parachute landing falls—are practiced again and again and again, so that the parachutist knows how best to absorb the shock of the fall. He also masters the use of the reserve chute on his chest, in case the main chute on his back somehow malfunctions. The training here, as in Basic, can be dull and monotonous and *life-saving*!

Finally, the day of the first jump came. I was nervous. All right, I was scared. When we loaded onto the aircraft in what are called "sticks," I really wondered, not if I'd ever get to the ground again, but whether I'd be alive to know it! Again I thought about volunteering; maybe this wasn't such a bright thing to do! I thought about a novel I had recently read called *The Man in the Grey Flannel Suit*, in which the hero kept telling himself "What the hell!" I'll admit I made my peace with God and then said to myself, "I'm ready." I was shocked as the jump-master went through his nine jump commands with lightning speed; I thought it would be a slow process! Just at the moment I was beginning to re-think the whole thing, I was out the door. My training took over, and I was counting, then looking up to check my canopy. I was airborne!

I managed to get through four more jumps without injury. In fact, in my entire airborne company there had been only a few injuries and they had all been minor. Thus I earned the silver wings of an Army paratrooper. I was prepared now to go to the "ports of embarkation." My assignment was USAREUR, the United States Army, Europe. Rebecca and I were married in Monson, Massachusetts on 13 September 1969. I wore my Army dress blue uniform, complete with shoulder boards with the gold bars and with the silver wings on my chest. Intellectually, I guess it didn't matter too much. But I was pretty proud of it. Looking back more than two decades later, I still am.

VII

That should have been the end of the airborne story, but there is a postscript. In 1990, twenty-one years after I graduated from airborne school, my son Chris, then a junior in Army ROTC at Notre Dame, went to Fort Benning to attend airborne school. His mother, younger brother, and I attended his graduation. It's customary in airborne school for the Master of Ceremonies at graduation to ask all second- or third-generation paratroopers to stand. When Chris stood up, I guess I felt a special sense of pride. I was able to pin his wings on. My eyes were just a little misty that day.

I did not use the old airborne "blood rite," in which the pinner pounds the wings with the heel of his hand, thus pushing the pointed tips into the flesh of the new paratrooper, drawing a little blood. I thought that was going too far. Plus, by that time, my son was a lot stronger than I.

Chris returned to Notre Dame that fall, and the Army ROTC there had a special ceremony for all airborne cadets. A Notre Dame distinguished alumnus had been invited to the ceremony to pin the wings onto Notre Dame's cadets. (Chris's mother and I were not at Notre Dame for this ceremony.) As he went through the line, pinning wings onto the BDUs of the cadets, he met Chris and saw the name Toner on the uniform. "Did you have an older brother [!] in OCS about twenty years ago?" he asked.

"No, sir," Chris said, "but my Dad was there."

"Jim Toner!" said the speaker.

He was Gene Cavanaugh, an old OCS classmate who remembered me, more than two decades after our graduation from Fort Benning. Such are the bonds forged by the OCS of the late 1960s. Gene did not know, and I am not sure whether Chris told him, that there was yet another connection Gene could not have known about: seven years after I graduated from OCS, I earned my Ph.D.—from Notre Dame.

NOTES

1. Robert A. Fitton, ed. *Leadership: Quotations from the Military Tradition* (Boulder, CO: Westview Press, 1990), p. 180.

2. [S.L.A. Marshall] *The Armed Forces Officer* (DOD Pam 1–20) (Washington, DC: Government Printing Office, 1960), p. 15. An excellent pamphlet-book that every officer should read. Because this is a government pamphlet, Marshall could not publicly be identified as the author.

3. On Calley, see Richard Hammer, *The Court-Martial of Lt. Calley* (New York: Coward, McGann & Geoghegan, Inc., 1971); and John Sack, *Lieutenant Calley: His Own Story* (New York: The Viking Press, 1971).

4. Two extraordinarily useful books that the junior leader should read are: Edward M. Flanagan, Jr., *Before the Battle* (Novato, CA: Presidio Press, 1985), a collection of what amounts to "tips" on practical leadership customs; and Dandridge M. Malone, *Small Unit Leadership: A Commonsense Approach* (Novato, CA: Presidio Press, 1983) in the same spirit as Flanagan's book. They are both of great potential use to the young officer.

5. As an example, Flanagan (above, n. 4) says that "a business, a work team, an individual must have an objective and a time-phased plan to achieve it" (p. 155).

6. See James MacGregor Burns, *Leadership* (New York: Harper Colophon, 1978), p. 18.

7. See, for example, John Keegan, *The Mask of Command* (New York: Viking, 1987), especially his imperative of "prescription," at pp. 318–321.

8. Larry Ingraham, "Fear and Loathing in the Barracks—And the Heart of Leadership," *Parameters* 18 (December 1988): 80.

9. General Lee is quoted in Fitton (above, n. 1), p. 92. Lord Nelson: "Duty—the sublimest word in the English language."

10. See *Currahee: The 506th Parachute Infantry Regiment, 20 July 1942–4 July 1945* (available from Fort Benning, Georgia).

3 SECOND LIEUTENANT

Whatever you are doing, put your whole heart into it.
—Colossians 3:23 (NEB)

The first responsibility of a leader is to define reality. The last is to say thank you. In between the two, the leader must become a servant and a debtor. That sums up the progress of an artful leader.
—Max De Pree, *Leadership is an Art*[1]

Someone once said that there are two tragedies. One is the failure of a man to realize his heart's fondest dream; the other is to achieve it.[2] All throughout my training, I feared, without any real foundation, that I would never obtain the dream of a commission. Now that I had the commission, and its responsibility, I hoped that I would not experience the other tragedy. I was willing to work very, very hard to be sure that I would bring honor to my commission. I had sworn that I would "support and defend the Constitution of the United States against all enemies, foreign and domestic, [and] that I [would] well and faithfully discharge the duties of the office upon which I [was] about to enter." I was being sent to Europe—a "peacetime" Army in time of war. I had heard that the problems there were severe. I was resolved, however, to give the job my level best, "so help me God." During my tour, I was to need every bit of the enthusiasm I marshaled in 1969 as I set off to take command of my platoon in the Federal Republic of Germany. The United States Army was in trouble, and that meant trouble for me, as it did for thousands of other young lieutenants in this, the "Age of Aquarius."

I

Ralph Waldo Emerson once said that nothing great was ever achieved without enthusiasm. Enthusiasm isn't a replacement for competence or fairness or effective organization (the three desiderata, in my judgment, of good leadership), but it is a necessary complement to the other three elements. Second lieutenants, with rare exceptions (for example, a former enlisted man commissioned through OCS or, occasionally, through ROTC or West Point), do not have experience; enthusiasm, therefore, is a practical must! Enthusiasm was one thing I had as I left Westover Air Force Base, Massachusetts, in October 1969 on the first leg of my trip to USAREUR, the United States Army, Europe. The first stop, a bit ironically, was Fort Dix, New Jersey, where I had arrived precisely one year earlier as a raw recruit to undergo basic combat training. Now I was back as an airborne infantry second lieutenant. After a short delay, I was to fly out of McGuire AFB en route to, well, I wasn't sure where. All I had was an APO—an Army Post Office (zip code 09175, which, later, I found out, was in Darmstadt, in Hesse).

But, for now, I returned to Fort Dix a conquering hero, a legend in my own mind. I exaggerate; but I did feel pretty good about my return there. A year ago I was a trainee, doing grass drills and dry shaving for a sadistic E-2. Now I was an officer, and I had earned my gold bars and silver wings the hard way. When I was married earlier in the month, one of my St. Anselm's College buddies (nicknamed Zelmo) had been in Monson, Massachusetts for the wedding. In conversation with him, I learned that he had been drafted and that he was going to Fort Dix for BCT. I promised, laughingly, to look him up there when I reported to Dix en route to Germany. In fact, I planned to redeem that pledge and, after getting my initial affairs settled, I set out on a hike to find my old friend Zelmo. I managed through the post locator to identify Zelmo's unit and, even though Dix was then in full swing, pumping out thousands of infantrymen, it didn't take me too long to find him. I knew the post pretty well from my earlier months there.

The walk into the basic training area brought back a flood of memories, a couple of them pleasant. It was after dark, and I had to search a little to find the right company. Finally, I found it and walked into the orderly room where I saw an NCO wearing the old Smokey-the-Bear campaign hat, which is still worn in the 1990s. I had strange, mixed emotions. The last NCOs I had dealt with for any appreciable amount of time were my airborne instructors. Before that (because my OCS "preceptors" had been tacs), my last contact with NCOs had been right here at Dix, where I had been under their complete authority. I somehow felt the urge to grovel before this drill sergeant! But I smiled inwardly, knowing that *he* was now subject to *me*! "Yes, sir! May I help you, sir?" he inquired.

God, I loved it! Being called "sir" by the airborne drill instructors hadn't meant much, for usually that respectful vocative was attached to an order commanding me to do pull-ups or push-ups. Now, somehow, it rang out, in all its simple majesty. "Drill sergeant," I replied, "I would like to see one of your trainees, Private Belmont."

"I am sorry, sir, but we prefer to keep the trainees separated from anyone who isn't in their company or cadre," the drill responded.

That was my start as an officer. The drill sergeant's refusal was also my first "challenge." I had made a simple, and not unfair, request of an NCO. For valid training reasons, he had respectfully declined my request. I saw myself bellowing, "Perkins [his last name], get your butt in gear *now* and have that trainee standing before me in zero-three minutes! That is a direct order." I just as quickly dismissed this Walter Mitty-like reverie. From 1969 to 1972, the period that I was on active duty as an officer, I never called an NCO by his last name without preceding it with "sergeant" or, when appropriate, "corporal." I thought that, if I expected my NCOs to comport themselves as officers without commissions (non-commissioned officers), I should always verbally award them their rank. I had learned in OCS that, according to the United States Army, there were two kinds of leaders, persuasive and authoritarian. Although there are times to be both, the good leader, I had been taught, led in a style most appropriate to his personality. I was more "persuasive" than "authoritarian." That was how I responded to this mild "challenge." "Drill Sergeant," I responded, "I know the rules here, and I understand your position. I'd like you to understand mine: Belmont and I were good friends in college; he was at my wedding; he's a draftee, probably on his way to Vietnam. I am off to USAREUR for a year. You and I both know that I may never see him again. I promised him I'd see him before I went overseas. Please help me to keep that promise." Had the sergeant refused, I am not sure what I would have done. I do know that I intended to see Zelmo, if need be by coming back the next day to talk with the BCT company commander. But it wasn't necessary.

The sergeant responded to my tone and temper in like manner. "O.K., sir, but could I ask you to keep it short?" In that very brief exchange, I confirmed what I had learned at Fort Benning. Normally, if I played it fair and square with my subordinates, without being either imperious or diffident, they responded well to me. So I got to see Zelmo that night. More than two decades later, I do not know what became of him.

II

Soon I was boarding a Military Airlift Command (MAC for short; soldiers joke that MAC means "Maybe Airplane Come") for the flight to Germany. When we landed and I deplaned, the woman customs official I had to deal with was about thirty years old, her blond hair tied severely back into a bun, with granny glasses, dressed very conservatively, with a stern and business-like manner. I thought, "The stereotype is true!" Well, it wasn't, but there are times I think that woman was planted there to reinforce it.

In a matter of minutes I was met by my two host officers, First Lieutenant Fred Wenton and First Lieutenant Don Dallas. They were two of the most incredible people I have ever known. The first was to DEROS (date eligible for return from overseas) out of USAREUR before too much longer and soon thereafter receive his honorable discharge from the Army, leaving behind him a reputation that probably to this day lives after him in USAREUR. The other was to die tragically not too long after the day he met me in Frankfurt; he died, in fact, the day after he was promoted to captain. The first was a military police officer and white; the second was an ordnance corps officer and black. The first was from Texas; the second, even with the Texas name, was from Illinois. They had become good friends. When they met me, they had no trouble recognizing me because I was in Army greens, with name tag. Wenton and Dallas were not only in civilian clothes, but in the most spectacular civilian clothes I had seen in a long time. Wenton dressed Texan; he looked as though he were about to go on a cattle drive. Dallas dressed, well, rococo, wearing the most ornate get-up I could imagine. Of course, at my most flamboyant, my best friends would call my dress "staid" or "stodgy." It was a sartorial indicator of things to come.

Wenton and Dallas settled me into the Bachelor Officer Quarters at Babenhausen, the home of the 36th Artillery Group, after telling me some very unsettling things about my unit, the 343rd Ordnance Company, located near Babenhausen. The Army usually makes assignments based on what is called a TO&E (a Table of Organization & Equipment). I had been assigned to the 343rd as an infantry officer; that much, they got right. But I was not to work in my MOS (military occupational specialty). I was to assume command of a platoon (that was O.K.—I had been trained exactly for that at OCS); but this was a "special weapons" platoon, or SASP. Now I had no idea whatsoever what was meant by special weapons, let alone by SASP. The next morning, after tossing and turning most of the night, I found out.

The trip to the 343rd lasted fifteen minutes. I reported to the commanding officer, Captain Lonnie R. Adams, an ordnance corps officer. "Sir, Lieutenant Toner reports," I said, saluting as I stood in front of the desk of my new C.O. The 343rd Ordnance Company was a mess. It didn't help that it was a bleak, chilly, overcast day. The company area was equally bleak; everything looked as if it needed a coat of paint. Usually, in the Army, things are painted. "If it moves, salute it; if it doesn't, pick it up; if you can't pick it up, paint it," goes the old adage. But here even the company headquarters—even the C.O.'s office and the orderly room—were ramshackle and rundown. I wasn't very impressed, but I thought I had better maintain my enthusiasm, which was waning rather rapidly, given the lugubrious accounts I had received the day before from Wenton and Dallas.

"Welcome to the 343rd, Lieutenant," said Captain Adams. "Have a seat."

"Thank you, sir," I responded, drawing up a chair. Adams proceeded to reinforce what his other lieutenants had told me in the process of briefing me about the 343rd. I was stunned. The 343rd had two SASPs—Special Ammunition Supply Points, which, in case of war, were to go to preassigned spots in the Federal Republic of Germany (FRG), there to dispense tactical nuclear weapons to certain U.S. Army units. In effect, I had just become a nuclear weapons officer, and I was to work in an MOS not even remotely similar to the one I had trained for at Dix and Benning. Good-by, infantry; good-by, airborne; hello, special weapons. The TO&E of a SASP was also entirely different from that of an infantry platoon. It was, in fact, similar to having my own company as a second lieutenant.

Captain Adams was a miniature battalion commander. But he was short of officers, of NCOs, of troops, of supplies. He was long on discipline problems, drug problems, and race problems. Training and morale were poor. Then it started to go downhill.

Like thousands of other Vietnam-era veterans, I had walked into the flip side of Vietnam. With the nation's attention, and military priorities, understandably riveted on southeast Asia, USAREUR was on the back burner for the Army. For the thousands of troops there, however, USAREUR was very important. It was where they did their tours, where many of them grew up, where some of them suffered, and where some of them died. This is not meant, in the least, to compare the hardships of USAREUR to those of Vietnam. In USAREUR we were not being shot at, ambushed, and killed by the enemy. The veteran of Vietnam deserves his memorial; the heartache is that it came so late for him. The sole suggestion here is that the USAREUR veteran served too; and most served well. With the very important exception that our lives were very rarely on the line in USAREUR, our problems there were often as bad as, and sometimes worse than, those of our fighting brothers in Vietnam.

The mess hall at the 343rd was in terrible shape. There was no EM club on post, and the NCO club (shared by the enlisted personnel) was a hazard—I mean a real engineering hazard. Because of its highly classified mission, the 343rd was isolated from all other posts; so the company commander was also the "installation coordinator." (At most Army posts, a number of companies or battalions or even higher elements share the same post facilities. Here the 343rd was the only game in town. There were no other units on the post or kaserne.) The company had its own chapel (a rusty metal hut with no means of warmth), its own motor pool (with scores of trucks that had to be maintained in the highest state of readiness possible because they might have to [and did] transport nuclear weapons—but there were very few real mechanics at the 343rd), its own post theater (badly rundown), its own barbershop (when its manager, a German national, came in), its own tiny PX (run by the wife of one of the NCOs on post), its own post office (run by troops given mail clerk duties), and its own education center (when teachers could be located).

In my limited Army experience, I had never heard of anything like this before! Neither had Captain Adams, who was brutally candid about the whole thing. "I just want to get out of here without getting relieved [that meant, in essence, getting fired, and endangering his career]"—a statement that did wonders for my morale! I was right back to wondering why I had joined the Army on my birthday. Then came the other shoe. "We have a real problem with drugs," said the beleaguered captain, "and with racism." Adams was a black officer, and he described to me how the post was badly split along racial lines. "I don't always get the cooperation I need from the white NCOs," Adams plaintively said. "To be frank, I know the black NCOs may not respond the way they should, either, to the white officers, most of whom are warrant officers." The post had four warrant officers (midway in the Army pecking order between NCOs and lieutenants.) On balance, I left the briefing disturbed and distressed. It got worse fast.

There hadn't been a platoon leader (a lieutenant) in the first SASP platoon for many months. Captain Adams introduced me to Staff Sergeant (E-6) Ned Colter, the platoon sergeant. Colter was a grizzled Korean War veteran, proud of the stripes he had earned there—with no promotion in the interim! At OCS I had been drilled on getting along well with the platoon sergeant. I had been told to listen carefully to his advice and counsel, all the time remembering that while I was technically in charge, I should make few moves until I learned the ropes from the non-commissioned officer in charge (NCOIC). Colter was not an educated man, but he was well schooled in the traditions of the Army, and I liked him. He was respectful and helpful. Then I learned a great truth, something overlooked in my OCS education. Staff Sergeant Colter could not wait to have me assume actual command of the troops, for that meant two things: (1) the responsibility would no longer be his; and (2) I would have to sign for all the platoon property, letting him off that hook. In time, I learned that Colter, although a veteran of twenty years of Army service, could not teach a number of military skills. He could not conduct training, for example, in map reading and in land navigation, critical to our operations in the German forests. He knew little about communications, about chemical/biological/radiological operations, and the like. Moreover, when it came to doing the platoon paperwork, Staff Sergeant Colter was lost. He was, on all these accounts, extremely happy to have a lieutenant. He was waiting for *my* advice. In fact, the situation was exactly opposite to what I had been led to expect from Infantry OCS. (I should add that, of the infantry NCOs I knew on post, a few of them were also poor map readers, while a handful of others were good at it.) One skill most frequently missing in the NCOs I encountered was a teaching or instructional skill, especially in a classroom or from a demonstration platform or stage. That meant problems, for we had a great deal of training to do. And at once.

III

One of the other things I noticed about the post was that the troops were spiritless and frequently disheveled. Their unkempt appearance was often ignored by the NCOs, who had learned to look the other way. "Looking the other way" became a trademark of the enlisted personnel, especially when an officer approached them. Instead of rendering a crisp salute and offering a verbal greeting, they would cross the street, if need be, to avoid saluting. I learned that black soldiers would often walk right past an officer and, if they saluted at all, it would be the raised, clenched fist of the black power salute, taunting the officer, whether black or white. The most militant blacks regarded black officers as "Uncle Toms" and white officers as "Massuh," the slave owner. The times, they were tough! I had already decided that a challenge of this sort could not be ignored. Someone once said that it is far better to anticipate than to react. So I began to plan strategy to deal with the myriad of problems I was beginning to see and to encounter.

My first plan was a failure. For years, the Army has insisted, understandably enough, on maintaining reasonably high standards of physical fitness. Today's Army, with its triad of running and push-ups and sit-ups, is probably the best-conditioned U.S. Army in history. Physical training can also be a great morale booster. Still an infantryman at heart, I seized upon the idea of a tough physical training (PT) regimen to build muscle and morale in the troops. I would lead them, setting the example. It was a Fort Benning school solution. But there is also this bit of doggerel to keep in mind with respect to such all-purpose fix-it ideas: "Here lie the bones of Lieutenant Jones, the pride of the institution. He died one night in his first firefight by applying the school solution." In our first platoon run, Staff Sergeant Colter was unable to jog more than a few dozen meters before he ignominiously fell out. My squad leaders, similarly, were terribly out of shape and completely unable to set the pace—even a slow pace—for their men. That was all right: only a handful of the men could run at anything like a PT-test pace. In my entire time in USAREUR, PT was never emphasized. Moreover, for the men of the 343rd, there was no running track or obstacle course or even gym of any sort.

Leadership always takes place, as I have written, "at the margin." A new platoon leader has only so much "currency" to spend, only so much slack his men will give him in war or peace before they dismiss him as a crank or incompetent. I was ready to yell "follow me," but hardly anyone was willing to accept the example at that point. As much as a vigorous PT program was clearly needed, I decided not to spend my money in that regard. I realized that, at the 343rd Ordnance Company, I was fighting the tide in pushing for PT, so I made a tactical retreat at once on that. Instead, I began to push, hard, on discipline, appearance, military courtesy, and camaraderie. The day after my failed jog, I called a platoon leaders' meeting. Looking back, I think it was one of the best things I did as a second lieutenant. In reality, I was on a kind of leadership reaction course, and I didn't know how to solve the problem; so I appealed to my comrades for help.

That evening, before a general platoon meeting I had called, I had a long talk with Staff Sergeant Colter and my squad leaders. I told them that the platoon needed more discipline and that we would set the example in all respects. I tried to be careful to let them know (although I will admit that I didn't really mean it) that they had done a good job, to this point, but that the absence of another helping hand (the lieutenant's) had, I said, hampered their efforts. I asked for their advice. To be candid, they had little to offer, for they were as dispirited as the troops, and it showed. But at least in asking them I had made them part of my plan. In outlining what I wanted to do, I did truly seek their counsel—something I had not done in my plan for the run—for I knew that if they didn't buy into my ideas and begin to model them, the troops never would. So their reactions were critical to what I said at the platoon meeting.

Everything we talked about in what I called the Leaders' Meeting (we were to have these once a week, or whenever any one of them wanted one, for they were strictly informal) was then condensed into a "point paper," a very concise handout that I gave to each NCO after our meeting. After checking it out with them, I would give copies to all of the troops after our general meeting the next day. The NCOs knew what was coming, so there would be no surprises for them. I continued to give out these "point papers" for several months until their contents became standard operating procedure for my platoon. The advantage of the paper was that it made platoon policy explicit. There was almost no chance of misunderstanding when troops could hear *and* read basic platoon policy.

At the general platoon meeting, I was appearing, actually, as platoon leader for the first time. I told the troops that our job was to prepare for war in Europe—a war that was unlikely to occur. It was unlikely to occur because we were there. We were our own best insurance policy. If war did occur—and Czechoslovakia had been invaded just "last" year—we had to be ready, or we would die. My job was to help them to be ready. I then took two chances that I suppose few old-time Army officers would agree with. First, I told them there would be no harassment or Mickey Mouse of any sort. That included organized PT. There would be no more runs. I knew that they knew I was infantry; I tried, therefore, to turn that to my advantage: "If any of you are thinking that I want this to be an airborne infantry unit, you're wrong. But I do want, and I expect, this to be a unit that we can be proud of. That means you do your job." They knew in a flash what I was saying: I would ignore some elements of Army life, provided I got full cooperation on others. My NCOs had agreed with the approach; it wasn't textbook, but I thought I had hit the right note.

Second, I told them that, *like them*, I was in the Army for a tour of duty, and *like them*, I intended to be as professional as possible during that tour so that in a couple of years when I returned to civilian life I would be proud of my service. I wanted to establish a tie to these soldiers without taking a cheap shot at the career NCOs among us, so I continued by saying that whether we were professional soldiers whose lives are dedicated to our country (very carefully chosen words) or soldiers in uniform for only a while (I avoided the phrase *short-term* or *short-time*, which had a special meaning [to me, indicative of apathy] in those days), we all wanted to do our duty the best we could. The way to do that was to work together.

I knew that this meeting was critical, so I had written down what I wanted the troops to take away from that meeting. I wanted them to know, first, that I was in charge; it was my meeting, not the sergeants' meeting (I had learned that lesson in basic and in OCS). I wanted them to know I was fair and reasonable. I was not going to run them to death—*but* they would perform creditably and conscientiously. Neither was I going to be a stickler for standards obtained nowhere else in our company or battalion—*but* that did not mean that we would not set an example. I told them I had no intention of insisting on spit-shined boots. "I have better things to do with my own time," I said. Nor did I care to see pressed fatigues, white sidewall (very close-cut) haircuts, or even highly glossed floors in the barracks. But a standard of neatness and cleanliness would be demanded. I put it in distinctly un-eloquent terms: "In other words," I said, "I'm not a jerk, but neither am I a fool." Before I distributed my point paper, I told them I knew that this wasn't basic training, and I had no intention of playing drill lieutenant. "You will find that if you deal squarely with me, I will do the same with you."

"There is another side to all this," I told them. Whenever and wherever I could, I would go to bat for them: promotions, letters of recommendation, comments for their files—all these would be routine in our platoon. In return, I expected that their duty performance would always be at the standard of their best reasonable effort. A sign was going to be put in front of our platoon: "Duty First." I told them that I wanted them to take that to heart and that I did not regard it as a football chant but as a soldier's motto. I adopted a calm, easy-going attitude, and I did not talk in the cadence or spit-fire way popular in military movies (or in BCT).

I had planned a final point. On my platoon leader door, I said, they would find a card (already placed there) stipulating the time each day, including Saturday and Sunday (for my first few months there), when there would be PT. That brought a low groan, and I know some of the guys in the platoon were thinking that I had already said we'd be reserved about PT. "No," I said, "not physical training—platoon time. If you have a problem, you can come to see me. I will take off or cover up this [lieutenant's] bar and we will talk man-to-man. It's not a bitch session, but if you have any problems with anything that I can help with, see me. If you have that opportunity and don't use it and still want to bitch, you're not only a lousy soldier but you're not much of a man, either." I ended by telling them that if we worked together, we'd beat a lot of problems, but if we didn't, the problems would beat a lot of us.

I know that the paragraphs above do not square precisely with standard advice given to green lieutenants. But leadership, beyond certain fundamentals common to virtually all situations, is the art of intelligent improvisation. A very successful Air Force junior officer might not, at least at first, be successful as a Marine Corps junior officer because he would find himself in circumstances very different from those for which he had been prepared. My infantry OCS training had simply not prepared me, specifically, for my assignment. USAREUR in 1969 was a melancholy place to be. Certain approaches to leadership, axiomatic in infantry platoons, would not have worked in my SASP platoon. I had to improvise, and I had to do it fast.

My open-door policy, for example, would have upset the NCOs, had they not known about it in advance. I promised them that the purpose of the policy was not to check on them; nor would I allow endless bitching about a given sergeant. I meant what I said, and I said what I meant. I wanted merely to take the pulse of the platoon—without crippling my chain of command. Two of my squad leaders were in the grade of specialist fifth class (a rank the Army no longer has). I told them that I intended to get them the stripes of a buck sergeant (same pay grade), something that was administratively permissible at that time. (I knew because I had already asked Captain Adams, and he had agreed to this lateral promotion.) My squad leaders would receive no more money, but they would have added prestige, and they enthusiastically agreed to my idea. The orders for my new sergeants were quick to appear, and I began addressing them immediately as "Sergeant Morgan" and "Sergeant Consolo."

The open-door policy, in my view, is not a sound one in a company in which there is strong leadership and an authority structure already in place. I knew I was in a military disaster area, and procedures that I employed at the 343rd might well be out of place in a well-run, tightly disciplined infantry company, with, presumably, a clearly competent C.O. (which I did not have) and with hard-driving, highly experienced NCOs (who were in very short supply at the 343rd). USAREUR, at that time, was in a state of siege, and officers were often afraid even to enter their own barracks after dark. That is why I improvised the way I did.

I told the NCOs that, if we expected the men to improve, we should be sure to set the example. Our own conduct and appearance should be beyond reproach. I told the NCOs that if they didn't respect their own rank, the troops never would. A few of my NCOs in this SASP platoon-company of mine lived in the barracks. (I was of a mind to move into the barracks myself, pending the arrival of my wife, in order to establish myself there; if my NCOs had not responded to my challenge, both my platoon sergeant and I would have been in that barracks.) Their rooms were barely cleaner and neater than those of their troops. I suggested that they set the example by having immaculate, organized rooms. They responded that they couldn't, for many cleaning supplies—and paint—were not made available to them. "Stay right here," I ordered.

I immediately sought out the supply sergeant and importuned him for all the supplies—and paint—we needed. The E-6 agreed with me, although he proved exceedingly difficult with my E-5 sergeants. So I was able to carry back at once the supplies needed by my NCOs.

Their rooms were stripped and cleaned and painted, as was my new office. I set up a platoon orderly room adjacent to my office, and I began holding the PT sessions as announced. The purpose of the platoon time, at first, was to let me interview each member of my platoon so that I would know each man personally. After the NCOs were set up, I told them that I wanted the men in their squads to emulate them, and that the first platoon inspection would be three days after they, the sergeants, had held their own formal inspection.

My first formal inspection was a terrific success. I told the platoon that, and I meant it. I then complimented my NCOs in front of the troops and complimented the platoon in front of the NCOs. In private, I gave each NCO a letter of commendation, the original of which I had sent to personnel for placement in their 201 files. That did it. The NCOs and I had worked out a modus operandi; daily inspections were their responsibility (although I insisted on occasional platoon leader inspections as well). Best of all, the NCOs were beginning to think of their squads as their men. I had started by doing their job for them; now they were beginning to do mine for me, and I was delighted.

The truth of the golden rule recurred constantly. I attempted to treat the men with obvious fairness and with a touch of humor. They responded to that, almost without exception. Whenever any one of them performed above the standards I set, I praised him orally, and in front of others. If someone failed, I used the sandwich technique. I counseled him privately, first telling him he had a good record or a good education or high intelligence (or whatever), then lodging my complaint and explaining it, then concluding with what I hoped would be a confidence builder about my expectation that he would help me build a solid platoon as a decent and workable place in which to fulfill our duties together. This bit of psychology was hardly remarkable in itself, but it was effective.

I tried to think through my policies before I announced them. I had been a private, so I knew enough to realize that what mattered was the private's perspective, not the lieutenant's. I truly kept asking myself, odd as it may sound here, if *I* worked for me, how would I respond to this or to that policy or requirement? My year of experience was working for me. I was, however, a second lieutenant, and I had a lot to learn. One day when I was entering the platoon orderly room, there were several people there, and Staff Sergeant Colter yelled, "Room, attention!" Sheepishly, I admit that I did what had been drilled into me for one good, solid year: I came to attention. I suppose it's a good thing that Sergeant Colter finally whispered, "Sir, it's for you," or we might all still be standing there. I felt like a fool. Of course, there was good reason for that feeling. All I could do was to smile and shake my head. I think that the people in the room had a good laugh at my expense. I did my best to laugh with them.

I was determined to do something for Staff Sergeant Colter, who was a good man. Not a gifted man, he was nonetheless a good soldier who proved loyal to me and efficient in carrying out my instructions. He also had a sense of humor, and we got along well. After several weeks of working with him, I couldn't understand why he wasn't an E-7 (platoon sergeant/sergeant first class). I decided to go to bat for him by writing an impassioned letter through the chain of command, seeking to help in the promotion process. I mentioned my intent at commander's call (the C.O.'s meeting) to a crusty CW4 by the name of Jim Michaels, who urged me not to send it. "Lieutenant, that's a mistake. Colter is bad news—he's a drinker." Now the warrant officers at the 343rd were not helicopter pilots, probably the most common warrant officers of the Vietnam era. They were technical specialists, ordnance experts, and they regarded themselves as general officers without the stars. They were good at what they did; they just were not anxious to pull duty. When officers had duties (such as officer of the day or OD) the warrants were quick to point out that they weren't commissioned officers. When the NCOs pulled duty (such as staff duty NCO), the warrants were quick to point out that they were, after all, officers.

In any case, when Mr. (or "Chief," by either of which warrants were called) Michaels lambasted my platoon sergeant, I came down hard on him. "That's crap," I replied. "Sergeant Colter is doing an exceptional job as my platoon sergeant, and he deserves E-7." Michaels just shook his head. He didn't sigh, "second lieutenants!" but he was thinking that, I know! I wrote that long letter for Staff Sergeant Colter, and I gave him a copy. He was happy. He was so happy that he went to the excuse we had for an NCO Club, got roaring drunk, drove his VW van wildly off post, was apprehended by German police, and shortly faced a DWI citation. My letter was returned to me.

"Sergeant Colter," I asked him later, using my great sandwich technique (I figured that if it were full of crow, I ought to eat it!), how could you do something so damned stupid?"

"I don't know, sir," he said. "Sometimes something just gets in my head, I guess." I am not sure, even today, what that meant, but something surely did get into his head. Michaels later just looked at—or rather right through—me.

IV

At the 343rd Ordnance Company, we had a back area where the nuclear weapons we had were stored in bunkers. A big inspection was coming up, and the company had to get the back area looking sharp in order to meet all the specifications. There was simply no way to accomplish that task in the time we had—in the normal workday time, that is. If we worked our tails off on a couple of Saturdays and Sundays and really busted during the regular week, we might be all right. We were woefully undermanned, but the nuclear weapons (and our inspectors) didn't care. The troops bought off on the Saturdays. We could order them, of course, to work, but the word *soldier* can mean to "make a pretense of working." If we used up too much of our currency with the troops, they would appear for work rather than face disciplinary charges, but would soldier on the job, or work in a cavalier and perfunctory manner, causing endless supervision problems. But the work had to be done, and the time was growing short. The troops of my platoon were responding well, but they were growing tired of cleaning and painting and policing (picking up trash) and putting up fences, and all the other things the job required. It was time to go the volunteer route. I went one weekend day to the ramshackle NCO Club, asking for volunteers to help me, saying that I, too, was a volunteer that day, and that I had come to work—not just to supervise. (Captain Adams, by the way, was nowhere in sight; nor, by that time, had I come to expect him to be. He knew nothing about setting an example.) All but one of the NCOs there responded to my call. We got the job done that day (and passed the inspection later) because we then got volunteer enlisted help who pitched in. I kept my word and worked right along with everyone else that day, even though I knew I wasn't going by the book. I did leave a little early, however, to get to the mess hall to be sure they stayed open late to accommodate our volunteers. I then cashed in all my chips with the mess steward and got him to serve ice cream and cake to the volunteers as a special, if very small, thank you. They also got sleep through privileges the next day.

The NCO who refused to help that day was Sergeant Meadows, a loser in every respect. I asked him in the NCO Club if he had anything special planned that day, and I would have accepted it if he had. "No,

I just don't get paid overtime and this is my day off." When I appealed to his pride in being an NCO, in working together with the troops, in setting a good example, and the like, he just snorted. That was it. Meadows, or so I had been told, had run down a Vietnamese with a jeep during his tour there; he had been busted to E-1 (why a mere wrist-slapping for what amounted to murder, I do not know). But he had it against the Army for demoting him! He had made his way back to E-5, but he intended to do nothing he didn't absolutely have to do. It was not to be my last problem with Meadows.

Officers are told, generally, not to do the work. They are paid to plan and supervise it. And, if they are working, they cannot do the overall supervisory tasks that must be accomplished. I threw all that to the winds that afternoon. I worked as hard as—or harder than—my troops that day. I got good and dirty. There is a time to be a pure supervisor and, admittedly, that is most of the time. There is a time, too, to lend a hand, to get filthy, to sweat—just as the troops are. It's a matter of timing to know the one from the other. This will sound Machiavellian, but I deliberately set out to seal my reputation that afternoon, and I think I did. Meadows had sealed his, too. I believe that every authority figure must seek out opportunities to build his image and reputation by means of calculated strategies. It shouldn't be phony. But looking for the place to seal a reputation is, in my judgment, sound policy. There is a story (that I can't verify) about General Patton who, while a cadet at West Point, suffered some injury, resulting in minor bleeding, during an activity; he refused all help until his task was completed, serving to establish a reputation. I was hardly a Patton! But I had read that account somewhere and remembered it; when I had the chance to apply it, I did. For example, I would occasionally come into my barracks—even when I wasn't OD—late at night, wearing spit-shined boots and crisp, starched fatigues, just waiting to be seen by a troop. It wasn't a practice I indulged in often, for it robbed me of sleep I needed. But the times I did it resulted, I think, in real dividends. When a troop saw me in the billets at, say 0300, he would of course be surprised. I would simply say, "Good morning, Jones. How are you?"

"Sir," Jones might say, "What are you doing here at this hour?"

"Just checking to make sure everything's O.K., Jones. How are you doing?" The idea was simple: to convey to the troops that I cared about them and their welfare. I did. But I had to seize every opportunity to show them I did. There are times and places to be a show-off. (Every good classroom teacher knows that!) If you show off what isn't real, you are perceived as a hypocrite—and rightly so. If you show off what is real, you are perceived as a leader—and rightly so.

The times were tough in the U.S. Army, Europe, of 1969 and beyond. If Army morale and discipline were beginning to decline with soldiers in Vietnam, these problems were even more advanced in Europe. Drug use was rampant. Racial episodes were common. Discipline was lax to the point of danger, for soldiers without discipline are an armed mob. Officers had enormous problems. My first two company commanders, Captain Adams and Captain McQueen, were summarily relieved. After McQueen's firing, First Lieutenant Don Dallas, the senior lieutenant in the company, took the reins of command. An ROTC officer with prior service as an EM, Dallas had been commissioned in the Ordnance Corps. He and his wife Ida lived in the apartment above Rebecca and me in the Babenhausen Housing Area. The Dallases had joined us in celebrating when the Military Airlift Command had flown Becky to join me just before Christmas 1969. In the three months I had known him, Don had seemed to be a pretty fair officer, and I was glad to see him in charge of the company.

Before too long, however, our real battalion commander had found an O-4, a Major Tann, to serve as company commander. Dallas needed more seasoning before having his own company. He looked forward to his promotion, to his first real command, and to an Army career. A few months later, I was sent on TDY (temporary duty) to an Army school to learn more about signal corps operations and communications equipment, for our company needed more expertise in that area than any of our officers or NCOs then had. The school was a few hours away from the 343rd. One day at the school, a lieutenant walking down the hall stopped me. "Hey, Toner, aren't you from the 343rd Ordnance?"

"Yeah," I replied nonchalantly.

"I just heard over AFN [Armed Forces Network radio] that an officer there dropped dead."

I called the company. It was true. Don Dallas had been promoted to captain on Friday afternoon. The battalion commander had shown up at company headquarters to pin on one set of the double silver bars of an Army captain onto his shoulder. Ida Dallas pinned the rank onto her husband's other shoulder. The next day, Dallas had shown up at the 343rd early in the morning to begin his shift as OD. It was about 0800, and he had gone upstairs to get a cup of coffee. He never made it back down under his own power. For reasons we never learned, Don Dallas dropped dead upstairs at the 343rd. It's possible there was still some life in him. The company medic was summoned, and I assume he administered CPR (or its 1969 version). They placed the dying Dallas into a company ambulance and took off at top speed for the 97th General Hospital in Frankfurt. It was clear en route that Captain Dallas was dead.

Major Tann and the company first sergeant had the duty of bringing the news to Ida Dallas, who was then nine months pregnant and expecting any day. Dallas's closest officer friend, Fred Wenton, had recently DEROSed back to the United States. Telling Ida was a terrible chore. She went into hysterics. She insisted upon seeing her husband's body, refusing to believe the reports of his death. Don's body had been taken to the morgue in the 97th General Hospital. The hospital seemed, in any case, a good place to bring Ida at that juncture, so the Major agreed to her entreaties. When she saw Don, she again went into hysterics. Less than twenty-four hours ago, she had been pinning captain's bars on him! His promotion party was to be held next Saturday night! It could not be! Ida wanted to go home to Chicago. No commercial airliner would take her on that long trip, given her condition at the time. With the help of the U.S. embassy, however, the red tape was cut, Army medical personnel and equipment went with her, and Ida made it safely to Chicago. Don and Ida's daughter was born there. All of their personal belongings and household goods were shipped to Ida as soon as possible thereafter. Don was twenty-four years old.

V

One of the commanders I worked for was a senior captain, George Oliver Perry Dolan, or Goppy for short. Captain Dolan was a West Pointer and an infantry officer. He was sent to the 343rd to maintain order and discipline, and he assumed command not long after Major Tann was sent to another command in a position more in keeping with his rank—although major was an appropriate rank for the company commander of the two SASPs. Dolan was from a military family, and he was desperate to make general. Goppy believed that his officers should be at work before him in the morning and leave after him at night. He was a very demanding C.O. I did not object to his stern, martinet-like ways. In some ways, I sympathized with him, for he had contracted malaria in Vietnam, and there were times when he was very ill. He was clearly a man of courage and conviction, if not particularly gifted. Leaders do not ply their profession, however, in a vacuum. Dolan had to understand that the 343rd was different from a hard-charging infantry line company. I tried to tell Dolan that in private but he would have none of it. I respected Dolan at first, but I certainly didn't like him. I also knew that it was far more important that he be respected than liked. The problem was that he was peremptory, cold, and seemingly even cruel to the troops. He was sent in to establish order, and, by God, order was what he was going to establish.

My job was to support Dolan and to be loyal to him. I believe I did both. A few months after he took command, he said to me—and he put it into an Officer Efficiency Report—that "Whenever I give a job to Lieutenant Toner, I can rest assured it will be done immediately and right." I took that, from him, as a supreme compliment. I think he respected me. But I don't think he liked me any better than I liked him. As the U.S. incursion into Cambodia began, Dolan and I were having a cup of coffee in the mess hall. I made a bad mistake in talking to him about anything other than company business, but I asked him for his opinion about the incursion. I shared with him my belief that the invasion of Cambodia made sound military sense, but then I raised the issue of what he would do if, say, he were commanded to lead an entirely unjustified invasion of Canada. Dolan apparently wanted to fight the War of 1812 all over again, insisting that he would love to have such an opportunity—and utterly missing the point. When I told him that I would refuse to follow such transparently unethical orders, he begin to fume. From that point on, Dolan and I seemed to go our opposite ways. Twenty years later, I had Dolan in mind when I began teaching a new course at the Air War College: "Command and Conscience."

Dolan infuriated me in small ways. He insisted that officers' boots be spit shined, that our fatigues be highly starched, that we wear a service ascot with our fatigues, that we come to work very early, and so on. Dolan was the commander, and I was serving as his executive officer. I thought some of his imperatives were Mickey Mouse, but I expressed that judgment to no one except my wife and Dolan himself, who by that time was hardly interested in my opinions. But what really bothered me was that he was asking more of the troops than those troops, at that time, were willing to give. He used up his currency on the wrong things! Dolan insisted on punctilious military courtesy. There was certainly a place for that, and I very strongly agreed—but not to excess. If he came into my office, for example, I was to snap to attention—"I mean snap!" He was a kind of Captain Queeg without the tiny silver balls. I urged him to lighten up just a little. I pleaded with him privately to insist upon the highest work standards but not to play Mickey Mouse with the troops, who were beginning to hate him, not because he was tough—that was all right—but because he was unnecessarily severe.

One afternoon, we had a commander's call. A good friend of mine, Lieutenant Binder, was among the twenty of us at the meeting. When Dolan came into the room, we snapped to attention, but Binder was tying his boot and was a second slow in assuming the position. Dolan just about went down Binder's throat, in front of the others (including NCOs), claiming that Binder was showing him no respect, that Binder was setting a lousy example, and so on. I was stunned. Not long after that, Goppy's wife was having a party for the wife of the new battalion commander. Becky was going because she enjoyed the socials, but Binder's wife Laurie chose not to go. Late one night, during pillow talk, Goppy's wife told him that Laurie Binder wasn't going to her party for the wife of Goppy's new boss. It must have been about 0100—very late in the Army! Goppy called Binder and ordered him to order his wife to attend that party. Dolan did not want to "look bad" in front of Mrs. Battalion Commander. Binder responded, "Sir, I am a soldier and you can order me to attend that party, and I will go. But my wife is not a soldier, and if she doesn't want to go, she doesn't have to. Good night, sir." And Binder hung up. In my book, he was right. The women's liberation movement must have been a terrible shock to Captain Dolan.

Binder was a dead officer—not literally, of course. But the next time Dolan had to do an OER, Binder got killed. Binder showed me a copy of the Officer Efficiency Report he had received from Dolan; it was a savage and grossly unfair assessment. Binder had the right to challenge it, and he asked me for my support in that attempt. I told him that I would of course help him. For example, Dolan had given Binder the lowest score possible on physical fitness. Binder had been an excellent college basketball player, and he was still in superb shape. I know. I had foolishly tried playing some hoops with him. Binder could have run circles around Dolan. The OER was purely vindictive. Binder asked all the other officers in the company to help him overturn that absurd OER. Every officer who was planning to leave the Army when his tour was over decided to help Binder. Every career-minded officer refused, saying that he was sorry, but, you know, first things first. I was sickened. Helping Binder for me, it is true, was a relatively easy thing to do, for I was not planning an Army career. The other guys, however, seemed gutless in their obsequious attitude toward Dolan and this travesty. I went to Dolan to report to him, man to man, that I intended to write a letter on behalf of Binder. Said Dolan: "I appreciate your telling me that, Lieutenant [every other time it was 'Jim']. At least I know where your loyalties lie." I started to interject something to the effect that I wasn't being disloyal to him—only loyal to the truth as I saw it, but he ended the conversation abruptly: "You're dismissed, Lieutenant."

"Sir," I began.

"You are *dismissed*, Lieutenant!"

I saluted, did an about-face, and left the office. Not too long after that episode, I was promoted to the rank of captain. Goppy came by to congratulate me. He told me that while he knew we were both captains, I should keep well in mind the fact that he was the senior captain, as well as the company commander. He expected me to act that way, as well as salute him outdoors, to set the proper example.

Goppy Dolan had a tough job to do. He wasn't well. He responded to a situation as his West Point training and his infantry preparation had taught him to. A courageous combat infantryman, Dolan was a patriot and, in many ways, a gentleman. He was also an ass. He had hardly any feel for the pulse of the men whom, supposedly, he was leading. Then, and now, I believe that Dolan saw all of us as tools to be used toward the achievement of his ultimate goal: general's stars. The American soldier can be led, in war or in peacetime. He can be pushed, too, but not too far (as Patton knew). The four most important words in pushing troops are *up to a point*. Dolan screwed around with the troops, and I know of not one incident in which he tempered justice with mercy. In the end, the troops got him. In Vietnam, they probably would have fragged him. In a peacetime Army, they killed him almost as surely, given his desperate dreams; they destroyed his career. Dolan had busted troops for riding in POVs (privately owned vehicles) without proper licenses. Somehow the troops learned that Goppy himself didn't have a POV license. One of the troops took a picture of Goppy driving his car and sent the photo to the MPs, asking them to check the vehicle registration and to check to see if Captain G. O. P. Dolan had a POV license. Not long after that, the battalion commander came to the 343rd to relieve Dolan. It was an ugly yelling scene, as Goppy no doubt realized that his career was in grave peril. He never asked me to help him challenge his relief or the OER he subsequently received. I would have refused, recalling the comment, in 1841, of the then Secretary of the Navy Abel P. Upshur: "He who finds it necessary to always exert his authority, lest those who are under him should forget that he possesses it, gives proof that he is deficient in the true dignity of character which impresses itself silently and without effort upon all who behold it."[3]

VI

The United States Army of 1969 was a conscription army. Due in part to the influence of the draft, the Army—probably more than was true of the other services—was filled with an incredible variety of people. In many ways, the Army is the most democratic of services,[4] because it draws its soldiers from all social strata and backgrounds. There were career soldiers (I use the term pejoratively on purpose), professional soldiers, and temporary soldiers. Soldiers in the last category could be subdivided into "professional temporaries" (soldiers who did their level best during their tour of duty) or "career temporaries" (by which oxymoron I refer to shirkers and malcontents). Goppy Dolan was a career soldier, despite his military family and West Point background. Later in the book, I offer a paean (see chapter 9, entitled "Rhapsody in Blue") to the professional soldier. Here I want to provide an example of the career temporary as well as of the professional temporary soldier.

Specialist Four (E-4) Richard DeLoom was from the Boston area. Having been born and raised in Massachusetts, I had a Bay State heritage that gave me something in common with this young man. To this day, I hope that was all I had in common with DeLoom, an extreme example of a "career temporary soldier." Whenever appropriate, I tried to get to know the soldiers of the 343rd on a personal basis, actively seeking them out and not just waiting for them to come to me during my PT. I tried equally hard, however, never to compromise my position in that respect. No enlisted personnel ever called me by my first name (to my face, at least). I came to know DeLoom too well. He was a draftee, and he hated the Army. He had already been sent to the stockade for theft, but he had been restored to active service. There were unsubstantiated rumors that DeLoom was selling Army equipment to the Germans for personal gain. I could not begin to understand why DeLoom had been assigned to the company supply room! As was true of many young troops of that day, DeLoom was not very well educated; he had dropped out of high school. But he was street smart and cunning. While in the stockade, he told me (and, to the extent I could check out the tale, I was able to confirm it), he had impregnated a German girl. Credulously, I asked him how that was possible. "Well," DeLoom responded, clearly taking pride in his accomplishment, "the stockade was surrounded by this fence, see? And the fence was broad-gauged chain-link, so my girl comes up real close to the fence . . . ." I got the picture.

One day, in talking to DeLoom (who fascinated me in a lurid kind of way), I decided to ask him, point-blank, about the barracks talk about him. "DeLoom," I began, "there are stories you're stealing the Army blind. There wouldn't be any truth in that now, would there?"

"Lieutenant," he responded, "I'm not saying it's true, of course, because I wasn't born yesterday. But I'll tell you this: if I could steal from the Army, I'd steal anything I could any time I could."

I had never thought of myself as Dr. Pangloss, thinking everything in the world was wonderful, but DeLoom shocked me. "Don't you think that you have an obligation to your country and to its Army to take care of the resources you are trusted with?" I started. Bishop Fulton J. Sheen or Dr. Billy Graham would have had nothing on me that day, as I began to wax eloquent about deontological ethics.

DeLoom was having none of it. "I got drafted," he said. "I didn't want to be here. My country ain't done nothin' for me 'cept to send me to the Army and then make me come to this hole. The way I look at it, I don't owe nobody nothin'."

I thought to myself, "How do I argue with that?" DeLoom would have done anything to anyone for any reason at any time—if there was something in it for him. In a sense, he was a moral idiot. The entire Army was a gigantic joke to DeLoom. But he did have a point in asking how the Army could be so stupid in putting him in supply and making him the clerk in charge of the Property Book, which is the unit's record of its possessions. In spite of myself, I really had to chuckle at that. The fox was in charge of the henhouse! We had a number of supply sergeants during the nearly three years of my tour at the 343rd. Some were pretty fair NCOs; others might have seen things DeLoom's way. When I read today about drug dealers who seem to be bereft of ethical sense, I think of DeLoom. I wonder, first, if he ever developed any sense of ethical responsibility; I wonder, second, if he is a multi-millionaire today; I wonder, third, if the two are mutually exclusive.

There is a parenthetical story I would like to relate about a junior NCO I also knew at the 343rd. He was Sergeant Gene Flannery. His case complements that of DeLoom. As a platoon leader, I had to counsel my people about what the Army calls "just debts." Some young men simply could not manage their finances, and I spent a number of hours trying to explain the rudiments of personal finance to them. I used to get very angry at some fly-by-night companies that hawked meretricious merchandise—flashy trash—to our troops. Beautiful jewelry was temporarily available to our troops at bargain rates—every pay day. The troops would buy the stuff on time, only to discover that their G.I. salaries hardly allowed the extra dollars necessary to buy that stuff. So they would fall behind in their payments. The rip-off companies would then send dunning letters through the soldier's chain of command, which we in the chain were expected to answer. (Someone apparently thought we had plenty of time for such things.) I lived up to that responsibility but usually attached caustic lines to the effect that "Well-managed companies offering high-quality merchandise seem to encounter few episodes of client payment delinquencies." Such obloquy probably wasn't called for, and perhaps I shouldn't have inserted it, but some of these overseas companies were little more than predators on young G.I.'s.

Flannery, in any case, was hardly guileless in his transactions. Although I found mitigating circumstances, frequently, in the cases of my soldiers who were not paying their bills, Flannery had a different and DeLoom-like view of things. He would buy dozens of things (cars, stereos, jewelry, books, etc.) without any intention whatever of making any payments. I used to get dozens—without exaggeration—of letters asking me to "counsel [him] about his financial obligations and the damage that can be done to his credit reputation as a result of not meeting those obligations." "Sergeant Flannery," I would, with a sigh, begin our counseling session, "you're behind in [say] your car payments. I have this letter from [x] asking me to state that you have been counseled about your obligations. Do you understand [once again] that this financial failure of yours to meet your obligations could be the basis of non-judicial punishment?"

"Yes, sir," Flannery would respond.

"Good. So are you going to make your payments, Sergeant Flannery?"

"No, sir."

"Sergeant, I don't understand."

"Sir, it's just a rip-off company anyway. I'll drive the car until they repossess it. Then I'll go get another one."

When I tried to explain to Flannery that no other car company would ever sell him another car, he rather condescendingly explained to me that companies would indeed sell him cars and that he had done this a number of times both in USAREUR and in CONUS (continental U. S.).

"But your credit rating!" I objected.

"They're so anxious to soak G.I.'s," he countered, "that they don't care about the credit ratings. They don't check 'em. They know they can repossess the car if they have to, and usually they get several payments because of command pressure [he looked right at me] before the G.I. can't meet the high payments. Then they get the car back anyway. I'm just playing their game."

There is a certain look that comedian Jack Benny made famous: that must have been my look at that moment. I would like to report that I had a snappy, Ann Landers-type of reply to all this and that somehow I managed to purify Flannery's philosophy of life. But I didn't. In fact, he made me feel like going out and buying a car on time, using it free until it was repossessed, playing a one-upmanship game on the avaricious predators who were exploiting G.I.'s. This is not to say that there weren't honest jewelry dealers, car sellers, insurance agents, mutual funds dealers, and other fair merchants in USAREUR; there certainly were. But Flannery, in his own way, had discovered how to beat the con artists (as long as we in the chain of command indulged him in his own deceptions and delinquencies). Do the ends justify the means? Do two wrongs make a right?

First Lieutenant Fred Wenton had first met me at the airport in Frankfurt (with Don Dallas). I had come to know this diminutive Texan MP officer very well. The TV series M*A*S*H had characters like Wenton. Nobody ever believes they're real; that was just a television show. Fred Wenton was real; and he was a "professional temporary" soldier or, if you prefer, a temporary professional. The 343rd demanded creative genius from its officers; Fred had an abundance of that. He was probably the most hilarious person I have ever known; yet he did his job with a taciturn commitment one rarely sees. Unlike DeLoom, Fred (who had no intention of remaining in the Army) considered himself to have a sense of social and military responsibility.[5] He had a pride of performance that was obvious, in an intuitive kind of way, to all who knew him. He took good care of his people, and he got his jobs done quickly and reliably. He was, at once, easy going and foul mouthed. He was the most unkempt officer I ever knew. He usually needed a haircut, he polished his low quarters (shoes) probably once (and that would have been under great duress when he was an ROTC cadet years before), and he wore unpressed fatigues too big for him. Yet he was respected by the troops because he was competent, and organized, and fair, and—in his own way—enthusiastic. And funny, very, very funny. Lieutenant Wenton thought that too much of the Army was gas—absurd, flatulent gas. Wenton never put on airs. What you saw was, as the expression goes, what you got.

Once a year, the battalions in the area held a formal ball, and officers (if they knew what was good for them and their OERs) went. Fred was not married, but it was expected that he would bring a date. I regret that Fred Wenton and Goppy Dolan never knew each other; their sole connection on the planet earth, as far as I know, is through me, for they were at the 343rd at different times. What a magical meeting it would have been between those two men! Wenton had been practically ordered by our battalion commander to put in an appearance at the ball. Wenton decided to go, which surprised me. At formal military balls, there is normally a receiving line. At this ball, the post commander of 36th Artillery Group (a full colonel) was at the end of the line, stiffly receiving, with his lady (wife), his subordinates, all of whom were sartorially resplendent in dress blues and black bow ties. Better late than never arrived First Lieutenant Fred Wenton. Time has dulled the exact details of his dress, but the spirit—with no hyperbole—of his "uniform" that night was as follows (and I am practically obliged to say here that Wenton was stone-cold sober): He was wearing one cowboy boot, one unpolished low quarter. He had his dress blue trousers tucked loosely into his cowboy boot and bloused over his low quarter. He wore a white dress shirt, with a black bow tie, but he also had some of his TA-50 gear (such as combat pack, a canteen, a first aid pouch) on over his dress white shirt. He topped it off with a helmet. Camouflage covered half his face. When he entered the Officers' Club, everything came to a dead halt. Like the television ad of recent years that had everyone straining to overhear some financial advice, with everything else coming to a screeching halt, when Wenton entered the club, nobody said anything. I heard a pin drop. On a carpet. I thought Wenton had lost his mind. He began going through the receiving line. Everyone in front of him obligingly cleared out, leaving him plenty of room. Officers in front of the receiving line had no idea what to say or to do; they reacted woodenly and shook hands with Wenton, their mouths agape. Only his voice could be heard: "Good evening, sir [or madam]. How do you do?" He approached the post commander. Would the colonel strike him? Would he have him arrested? Would he lecture him and all of us about wearing the uniform?

The colonel laughed. Like some gigantic row of dominoes, everyone else began to laugh. Fred, having now removed his steel pot, chatted amiably with the colonel, and the evening's festivities went on. I don't think the colonel, confronted with the magnificent absurdity of it all, could have helped himself. He had to laugh.

Fred later contended that the colonel was O.K., and not full of gas. Fred thought that most senior Army officers were pompous asses. He recounted the story to me (I checked it with two NCOs, who corroborated it) of what happened once to one high-ranking officer from the Fifth U.S. Corps in Germany who was inspecting Fred's platoon's barracks, which were old and rundown. This Army colonel had the usual retinue of fawning servants with him that all high-ranking officers seem to have. (Why is it that a general officer has to have his own aide, plus a battery of secretaries, drivers, and other assorted personnel?) Fred, being Fred, met the officer with a barrage of complaints about the barracks, especially about engineering problems over which he had no control. Fred was always more the inspector than the inspected. He complained relentlessly about the plumbing, about the electricity, about the roofing—all things he himself could not improve, all the time asking the colonel what he intended to do about it! As the colonel got on with his perfunctory inspection, Fred kept up the fire, demanding to know why the 343rd was not getting the help it needed from higher headquarters. The colonel's sycophantic entourage was no doubt scandalized by this irreverent first lieutenant. The colonel went into the latrine and found there, in a commode, to his horror and dismay, some human excrement floating in the water. "What's this, lieutenant?" he bellowed.

"That's shit, sir," Wenton answered, straight-faced.

Up and down the long retinue of the colonel's lieutenant colonels, majors, captains, lieutenants, sergeants, and one PFC (pen and notebook in hand) went the burden of the colonel's discovery: "Shit in the commode!"

"Why wasn't this flushed, lieutenant?" the colonel asked. Up and down the length of the retinue echoed that timeless question, "Why wasn't it flushed?" The PFC recorded the question for posterity.

"Sir," Fred explained, "Most of the commodes just don't work, and I wanted you to see that, not just gaze at fresh water. We need some engineering help here, sir." No echo on that.

"Well, we'll just see about this commode," said the colonel.

"The colonel's going to see about the commode" was now the repeated shibboleth.

"All right, sir," said Fred, "But it's not just stopped up. The whole piping system is shot," and with that he stepped out of the colonel's stall, leaving the colonel, the retinue and the excrement to their summit conference. The colonel flushed the commode.

"The colonel's flushing the commode" went the chant. All of the retinue crowded in, attempting to share in the vision of the colonel's expected plumbing triumph. The water rose. And rose. The colonel, now crowded into the stall by the gaping entourage at his rear, suddenly realized that Wenton had been right. The toilet didn't work, and it was about to flood over, right at him.

"Back up! Let me out!" ordered the colonel in his best command voice.

All up and down the retinue the mantra was repeated: "Back up! Let the colonel out!"

Too late. By the time the PFC at the back of the line was recording this historic command, the colonel, pressed into the stall by his avid followers, watched in horror as the excrement flooded over the top, most of it landing squarely on the colonel's brightly polished shoes. It was more than Wenton and his platoon sergeant could bear, as they struggled manfully to control their maniacal laughter. By now most of the retinue had retreated, allowing their colonel, in his turn, to stage a withdrawal from the flooding commode, and Lieutenant Wenton had time enough to gain a measure of self-control. It didn't matter too much. The inspection was over.

As an MP officer, Lieutenant Wenton had to coordinate the storage efforts there of a number of battalions whose special weapons were stored by the 343rd Ordnance Company. Occasionally, a general officer would come by for a pleasant visit, expecting to be briefed by the OIC (Officer in Charge) of our back area, who was Lieutenant Wenton. I was always amazed as a junior officer how senior officers expected their drones to have the time, energy, and resources to prepare "briefings." There wasn't time for such showmanship. Wenton never said to such touring officers, "Look, we haven't got the men, money, or materials to tell you, in thirty minutes, complete with color flip charts, what you need—really need—to find out for yourselves. Come out here and work with us for two days. Get your boots in the mud. Pull a little duty with the troops; eat with them; sleep with them; sweat with them. Remember for a day or two what it was like being a lieutenant without a bunch of bootlickers following you around. Then you won't need your damned 'briefing.'" But he should have, for we all felt that way. But Fred was honest. One afternoon, he had to brief a general on the back area, where our nuclear weapons were stored and secured. A number of battalion commanders, mostly lieutenant colonels, were there for the briefing, too, because it involved weapons assigned to them in the event of war.

Now Fred had been having trouble with one lieutenant colonel who constantly ignored Fred's legitimate requests for work to be done on and around the storage bunkers assigned to his battalion. That battalion commander had constantly pulled rank on Fred, ordering Fred to do the work that his battalion was actually supposed to do. Bad mistake. Fred went on in his usual way in the briefing, except that he refused to use flip charts, slides, overlays, and the rest, contending that these things were gas. The general asked the usual and customary—and perfunctory—questions. Fred waited for the general's routine question about cooperation, and then he unloaded. In the Army, with very rare exceptions, nobody ever dumps on anyone else (when that person is there) in front of a general. It's simple logrolling: you tell the general I'm O.K., and I tell the general you're O.K. Not Fred. He seized the moment, telling the general that "For the most part, sir, we get good cooperation here from all the battalions that use this facility [the assembled group of battalion commanders smiled and nodded approvingly]—except for Lieutenant Colonel Williams [Williams's mouth dropped open; the other colonels' mouths dropped open, too]. Sir, I have repeatedly asked him to paint and clean his bunker, to repair the fences, to mow the grass in the cleared area around it, to supply a fair number of trained guards, and so on. All I have gotten from Lieutenant Colonel Williams is excuses, rank-pulling, and lack of consideration. You asked me a question, sir, and you have your answer!"

The general was doubtless as shocked by this candor as were his underlings. But now the ball was in his court, and he couldn't lose face by ignoring what appeared to be dereliction of duty by one of his colonels. "Thank you, lieutenant. Your frankness is appreciated. We will look into this matter at once." Lieutenant Colonel Williams, his career dissolving in front of him, was about ready to spring at Wenton, murder in his heart and in his eyes. Wenton smiled benevolently at the assembled group. The next time Fred asked any of the colonels at that meeting for help *in anything*, he had it—at once.

A visiting four-star general toured the 343rd one afternoon in a whirlwind inspection. To us lieutenants, such an inspection was stupid. Everyone knew he was coming; everyone dressed up and polished and shined and put on a show, taking hundreds of man hours from our real jobs. But captains' careers were made or broken by the impressions that their lieutenant colonels had of them, and so the boots were shined. Lieutenant colonels had their careers made or broken by the impressions that the full colonels or generals had of them, and so the brass was polished to gleaming. Far be it for the inspectors to see real troops at real work with real dirt or real sweat on their fatigues. Everything became a Potemkin Village, a contrived and managed image, with all rocks painted and all trees painted (but only half-way up the trunk). Meanwhile, the troops knew that the whole thing was a joke. While they scrubbed and polished and cleaned and washed and waxed and shined and buffed so that their officers would look good, the real work lay undone. And generals too often ignored the real work, for that couldn't be painted or briefed. Fred was generally disgusted with this typical Army nonsense, and when the general departed that afternoon, as dozens of assembled military dignitaries stood ramrod straight, holding their salutes and smiling toward the general's helicopter, Fred stood, hat in hand, waving it up and down, and yelling, "Bye, bye, General; bye, bye, General. Come again soon!" The inanity of it all was infectious and some of us, once again incredulous at Wenton's antics, couldn't resist laughing—much to the discomfiture of our battalion commander.

When Wenton left the 343rd to return to CONUS and to be honorably discharged from the Army, I knew I would miss him. More to the point, so would the 343rd. A very colorful character, Wenton could outrage traditionalists of every sort (including me, at least with respect to some of his notions). But he accomplished his missions in a time of deep trouble for the Army. He truly stood up for his men, and he cared for them. He brought efficiency and (maybe too much) humor to a thankless job, to a virtually forgotten and abandoned post, in bleak times. He was a Vietnam-era veteran who had done a soldier's job for his country. No one, I am sure, ever thanked him. Until now. Thanks for the memories, Fred. You were a good soldier and a brother officer.

NOTES

1. Max De Pree, *Leadership is an Art* (New York: Doubleday, 1989), p. 9.

2. In *Man and Superman* (1903, various publishers), George Bernard Shaw wrote that "There are two tragedies in life. One is to lose your heart's desire. The other is to gain it" (Act IV). Oscar Wilde, in *Lady Windermere's Fan* (1892), had put it this way: "In this world there are only two tragedies. One is not getting what one wants, and the other is getting it" (Act III).

3. Quoted in Henry E. Eccles, *Military Power in a Free Society* (Newport, RI: Naval War College Press, 1979), p. 195.

4. For an analysis of "American Military Styles in Strategy and Analysis," see Carl H. Builder, *The Masks of War* (Baltimore, MD: The Johns Hopkins University Press, 1989).

5. Two worthwhile books about military leadership are Aubrey S. Newman, *Follow Me: The Human Element in Leadership* (Novato, CA: Presidio Press, 1981) and W. J. Wood, *Leaders and Battles: The Art of Military Leadership* (Novato, CA: Presidio Press, 1984), which studies courage, will, intellect, presence, and energy. Those qualities are necessary in combat leaders; they are also necessary in garrison leaders.

4 FIRST LIEUTENANT

Set and demand standards of excellence. Anybody who accepts mediocrity—in school, on the job, in life—is a guy who compromises. And when the leader compromises, the whole damn organization compromises.

—Charles Knight, in *Time*, 25 February 1980

The badge of rank which an officer wears on his coat is really a symbol of servitude to his men.

—General Maxwell Taylor, *The Field Artillery Journal*, Jan/Feb 1947

The strength of the leader is the strength of the group.

—Vince Lombardi

I

One of the books I bought before graduating from OCS, *The Officer's Guide*, was a useful compilation of advice about a myriad of things, some of them strange and new, in the life of a young lieutenant. One of the bits of guidance to be found there was this: "Personal integrity is the foundation upon which the standing of an officer rests in the minds of subordinates, associates, and superior officers. It is simple, basic, personal honesty."[1] An abundance of testimony exists about the need for integrity among combat soldiers. Imagine a patrol leader too lazy or too cowardly to recon an area but who reports the area all clear, leading, possibly, to another patrol's being ambushed later in that same area because of their belief in the honesty of the report they received earlier. But there is no less need for integrity in the peacetime Army. I have never been certain whether peacetime is the right word to apply to the United States Army, Europe in the early 1970s; in a way, we were in a state of siege, assaulted on all sides by the evils of the day. I did know that I had ample opportunity to betray the standards I had learned as a child from my parents, as a student in high school and college, and as an officer candidate. Like the word *pride*, the word *compromise* can have both positive and negative connotations. Charles Knight (see the preceding epigraph) notwithstanding, there is a time to compromise. (Without the spirit of compromise, the founding fathers in Philadelphia in 1787 could not have produced our constitution, a "bundle of compromises.") What Knight means by compromise, of course, is to "sell out" or to sacrifice all honor or justice on the altar of private gain. In the same sense, having "pride," the dictionary says, can mean "too high an opinion of one's importance or superiority" *or* "dignified self-respect." The lieutenant had better be able to distinguish between types of pride and types of compromise. I wasn't always able to make the right differentiations.

If the story of Fred Wenton ended, on balance, fairly happily, the story of Staff Sergeant Colter, my platoon sergeant, ended sadly. After his drunk driving mess, Staff Sergeant Colter seemed to realize that his prospects of making E-7 were non-existent. He fell into a depression. I was aware of his badly flagging spirits and sought to counsel him. He was in the wrong place at the wrong time. It was possible, at this time, for a young man right out of high school to enlist in the Army, to go through basic and AIT, and then to go to "instant NCO" school at Fort Benning. Similar to OCS, the instant NCO course offered certain troops the chance to earn the stripes of an E-5 sergeant after just a few weeks at the NCO academy. Usually the top graduate of the course was made an E-6 (or staff sergeant).

Somehow, Colter didn't mind that OCS graduates became lieutenants after six months. In his mind, they were all college graduates, although of course that was not true. But that a young "kid" could wear his proud sergeant chevrons after only a few months in the Army—it was all too much for him, especially after the 343rd received a shipment of instant NCOs. My experience with them, as with OCS, West Point, and ROTC officers, was mixed. Some were excellent. Some simply didn't have enough command presence, stuff, sense of urgency, or drive to take charge of other young men their own age and get them to do what they were supposed to. I was young, too, at twenty-four (as a first lieutenant [as of 8 August 1970, one year after OCS]), but I had the gold or silver bars of a lieutenant to insulate me from the men if and when I had to make an unpopular decision.

That is one of the reasons, and it's a good one, for maintaining a proper senior-subordinate distance and appropriate aloofness. The new buck sergeants didn't have that kind of social distance from their charges, and it was tough for them to know where familiarity ended and insistence on respect began. Zany as Lieutenant Wenton had been, none of his men had ever addressed him by his first name—or failed to take his directions to heart. Our officers were not addressed as "L-T," although the vocative "lieutenant" was sometimes used. Good sergeants were not called "Sarge"; they were sergeants. The young instant NCOs had to learn how to be friendly without being friends—not an easy thing for them (or for some officers or teachers or doctors) to learn. The man or woman in authority must keep constantly in mind that the authority exercised and the responsibility borne coincide always with the willingness of that leader, first of all, to be serious about his or her office. If the leader mocks his own office, or refuses to be serious about it, he must not be surprised when his subordinates do. The instants NCOs, or some of them, hadn't yet apprehended that simple fact. For all his shenanigans, Wenton carefully chose his times and places to dilute his dignity. The instant NCOs, in living with the troops, had understandable difficulty in putting on, as Keegan calls it, "the mask of command."[2] If no man is a hero to his own butler, neither is any NCO an authority figure to his troops unless (1) he insists upon it at the proper times, and (2) they are recognized as being competent, organized, fair, and enthusiastic.

Staff Sergeant Colter never had to deal with these matters. He did not live in the barracks. He was an old-time NCO. And he could not understand "these kids of today," as he called them. It was time for Colter to take his retirement and head home to South Carolina. I knew it and so did he. Unless someone has been a troop commander, in war or peace, he will not understand how some old soldiers deeply love the Army. It is, in a very real sense, their life. For the old NCOs, the 1960s and 1970s were a living hell. The troops, with their drugs, acid rock music, long hair, and anti-patriotic attitude disgusted and confused the old sergeants. The adage that NCOs are the backbone of the Army is undeniably true. Officers may set the course, and may have to, but it is the foremen of the company, the NCOs, who see that the task is understood and supervised and accomplished. NCOs, as I had learned in basic training, work for a living. By the way, as much as an NCO will appreciate an officer's willingness to get down and dirty with the troops, that should be a rarity, for that is not the officer's job, as the NCO will be quick to point out. Officers must be very careful never to undermine the authority, or to undercut in any way the dignity, of their non-commissioned officers. All these things Colter knew, even if he couldn't articulate them. As the Army turned volunteer in the early 1970s, the idea of an NCO taking a recalcitrant troop out behind the barracks and "changing his mind" about a few things became, largely, a thing of the past. The world had done a mean thing to Sergeant Colter: it had changed.

One afternoon, my company commander, at that time Major Tann, received a call from the Babenhausen housing area officer in charge. It involved Colter. In a matter of minutes, Tann had summoned me to his office. "Lieutenant, your platoon sergeant is drunk in his apartment, making a racket, and threatening to kill his wife with a shotgun if anybody bothers him in any way." That was pretty bad. But what came next was worse: "You're going to have to do something about it." I wondered where in my OCS notebooks were my jottings about how to deal with this kind of problem.

Tann and I rode out to the housing area. During the short ride, I reviewed in my mind what I knew about domestic crisis intervention. My complete review didn't take me long: I didn't know very much. When we got to the housing area, the major told me to "take that damned shotgun away from your platoon sergeant" and to "turn him over to the MPs." He did not suggest a specific plan to me in order to accomplish that mission. I saw the headline in the next day's *Stars and Stripes*: "Lieutenant Shotgunned to Death by Platoon Sergeant." The drama reached its acme when Major Tann told me he'd be waiting for me in the car. Terrific. I found the MP lieutenant and told him, "Get your men out of here! I don't want Colter to see them. If he shoots his wife [or me, I was also thinking], they'll have to go get him. Otherwise, keep a very low profile until he comes out. Come in only if you hear shots." The building had been evacuated so, as far as the MPs could determine, it would be Colter, Mrs. Colter, and me—with Major Tann in reserve. In his car.

In the movies, the guy who gets this kind of assignment seems suave and self-assured. I was scared half to death. I really thought I'd manage to make it the rest of the way to death, courtesy of Colter's shotgun. I climbed the stairs to Colter's apartment. Moving to the side of the door and truly expecting a shotgun blast to rip the door apart, I knocked twice on the door while announcing, "Sergeant Colter, it's me, Lieutenant Toner. Can I talk to you?" I had chosen not to carry a pistol, for I wanted Colter to see that I was unarmed. Major Tann had a .45 with him, and the MPs were loaded for bear. I was a little ashamed of the pitch of my voice as I asked Colter if I could talk with him. It went very high on me, indicating, as if I didn't already know it, the advanced stage of fear I was feeling. I was supposed to be an infantry officer, but thinking of Colter's shotgun was trouble enough for me. This single, simple episode has always made me appreciate the enormous courage shown by soldiers facing real fire from a real enemy. I was greatly relieved to hear Colter's voice come through the door. "Oh, Lieutenant Toner, come on in, sir," he said in a most agreeable and mellifluous voice. I couldn't go anywhere; the door was locked. All that for nothing.

"I can't get in, Sergeant Colter. The door is locked."

"Sorry, sir. Just a minute." Colter came to the door, shotgun in hand. His appearance shocked me—unshaven, wearing an old undershirt, obviously drunk. "Come on in. Want a drink?"

My mind flashed to what I "ought" to do. Accept the drink to prove my friendliness? Decline it to set an example? I realized that I had no idea what I "ought" to do. "No, thanks, Sergeant Colter." I looked at his wife, who had been beaten. She sat, I think more ashamed than afraid or angry, on a wooden chair in the middle of the living room. She and Colter were proud hillbilly folk (not a pejorative term at all, by the way). Others, principally me, were now involved in their domestic dispute. Her shame and mortification became clear, although she had done nothing to warrant this situation. "Look, Sergeant Colter, that weapon's making me a little nervous. Why don't you just put it down carefully or give it to me?"

He surprised me by saying, "Sure, sir—but only if you'll have that drink with me." I drink an occasional beer, but I have never liked alcohol; that day, however, I'd have drunk a case of whiskey to get that weapon away from Colter. I accepted his offer at once. He gave me the gun. Mrs. Colter sighed. She was all right, if badly shaken. I had the drink with Sergeant Colter. Then we walked out together to the MPs. Others went in to tend to Mrs. Colter.

I felt terrible for Mrs. Colter, a proud mountain woman whose "shame" was thus advertised. I felt terrible, too, for her husband. Of course there is no justification whatever for what he did. His penalty was, in a way, light. His retirement process was instituted immediately. Because of his record and because his wife pressed no charges, he "got off." He returned to duty long enough to outprocess, his Army career—to him his life—ended. Still, that was not the end. I talked with Colter when he came back to the platoon. I told him that I needed his advice and counsel in the few days he had left as he cleared post, waited for his final orders to come, and arranged for the packing of his household goods. I told him I hoped that the new platoon sergeant would arrive in time to be briefed by Colter before he left. I tried to keep him active while he was on post but, when duty hours were over, he went to the NCO Club. I had talked the matter through with the company first sergeant. "Top, we have to keep someone on Colter. He's going to drink. We both know that. So I want an NCO to keep tabs on him, to drive him home, and to keep him squared away until he leaves." The plan worked for a couple of days. The first sergeant set up a very informal duty roster of the NCOs who could and would watch out for Colter without his knowing about it. On the third or fourth day before his orders came in, he left the club, drunk, while his escort was in the latrine.

Our back-up plan failed, too; for even though I had left instructions with the post gateguard that Colter was not to leave post without an escort, the guard waved him through, not knowing the orders. I blame myself for that. I had not wanted to put such an order in writing or even to announce it publicly to the guard staff. I had let it go pretty much by word of mouth—a mistake. The result was that few guards got the message. In trying to save Colter humiliation and in trying to watch out for his dignity, I had compromised in a bad way; I had failed in my obligation to keep him on post unless escorted. He sailed off post, a drunken driver. He made it part way to Babenhausen before his VW van drove straight into a tree. Thank God, Colter hurt only himself that day. He didn't careen into any other car or into a pedestrian. I got word of the accident and took off in my car to find him. By the time I got there, the medics were bringing him to the 97th General Hospital in Frankfurt. It looked bad, blood everywhere. Poor Mrs. Colter! After everything, she was still loyal to her soldier-husband; they had been through everything together. Colter survived, although in the crash he had received a haircut that looked as though he had been scalped, and what couldn't be seen because of the bandages was probably worse. Colter was sent to the U.S. directly from the hospital, to be retired. It was a sad and bitter end for an old soldier and the wife who loved him. No instant NCO I ever knew got himself and his wife into such a mess.

II

Leadership, of course, is critical to every human enterprise. In its academies, in its ROTC programs, and in its OCS programs, the military services have tried over the years to inculcate into its rising leaders the standards expected of them and the standards, in turn, that those newly commissioned leaders are to expect of their charges. If the Pope or the bishops were to prove incompetent or unethical (the argument will be put forward later in the book that the two failings are symbiotic), the body of priests would likely soon follow suit. That is the way of a hierarchy. Effective leadership is a demonstration of excellence, not merely a clarion call for it. Good senior leaders can imbue junior leaders with requisite skills and necessary attitudes. Max De Pree put this idea succinctly recently: "The measure of leadership is not the quality of the head [the boss], but the tone of the followers [the troops or students or workers]. The signs of outstanding leadership appear primarily among the followers. Are the followers reaching their potential? Are they learning? Serving? Do they achieve the required results? Do they change with grace? Manage conflict?"[3]

But there is always a flip side. Good teachers can have failing students. Jesus had Judas. Socrates had Alcibiades. Good senior officers can have failing junior officers. The commander, it is true, is responsible for all that his command does or fails to do. But sometimes that responsibility is very indirect; if the commander has complete responsibility, he had better have concomitant absolute authority. And nobody wants that. There is such a thing as individual responsibility, whether it involves Judas or Calley—or Second Lieutenant John A. Bolling at the 343rd Ordnance Company.

Lieutenant Bolling, while no Lieutenant Calley, may have been beyond military redemption. Bolling was an oddity in that he was, well, a hippie OCS graduate. ROTC is understandably more tolerant of diversity among its college student-cadets than active-duty Officer Candidate Schools are (or should be). Bolling was as close to being a hippie as I ever saw during my Army tour. Once he enrolled in OCS, like many of the rest of us, Bolling had the stamina to finish. He was a bright and energetic young college graduate.

He had little use for the Army, however, and was not above making his militarily iconoclastic views known—even to the troops. I think that was his way of cultivating popularity with his troops. Bolling and I shared a similar view of some of the incompetents we had to deal with, particularly on the battalion staff. There is a vital difference, however, between two lieutenants complaining between themselves about the sometimes little-better-than-moronic practices they have to contend with and the same sentiments expressed to the troops; to do that is unprofessional in the extreme. Bolling constantly tried to be one of the boys with his troops (forsaking his office), causing enormous (if tacit) resentment from his platoon sergeant, a professional NCO whose boss, after all, was Lieutenant Bolling. Although Bolling and I got along, we had very different approaches to things. He was a bachelor who drove a fast sports car. I was married with a VW beetle. And those were only two indicators of different approaches to the problems that Bolling and I faced at the 343rd.

Because our job was the security of nuclear weapons, we had to undergo "surety testing." Our records were flagged so that if we showed evidence of alcohol abuse, drug use, homosexuality, or of other conduct or affections deemed contrary to the interests of military security, we could be summarily removed from the 343rd. The surety testing wasn't very sure: consider the case of Staff Sergeant Colter! I had reason to believe that Lieutenant Bolling was using drugs. But I had no more than suspicions. I had no evidence in that respect, and nobody ever asked me about my beliefs concerning Bolling's possible use of narcotics.

He and I were friends, and I was increasingly bothered by what I saw as a contradiction between his lifestyle and my military obligation with regard to surety. I went to his office one afternoon to talk it out. After ensuring our privacy, I said, "John, you have to know something. I do not know for sure whether you use drugs or not. I don't want to know because it's none of my business. [I suppose, however, that it was.] But if I discover that you are using drugs, I want you to know that I'll report that immediately to the Battalion S-2 [intelligence officer]."

Whether or not Bolling saw me as a brother resister against the establishment, I do not know. He knew that I regarded some Army practices as counterproductive at best and some of our superiors as inept; he knew I was not planning an Army career. He may have drawn the wrong conclusions. After my admonition, he was quiet for a few moments. He finally replied, "Jim, I understand that. There are no problems!" Maybe none had ever existed. Maybe my holier-than-thou approach was not necessary in the least, causing needless resentment. Maybe I was pursuing where none were fleeing. But I thought that I should deal with the possible problem—anticipation rather than reaction and all that. I never learned of any drug connection Bolling had, and it soon became academic anyway. He transferred out, I believe of his own volition, to a less sensitive unit.

I was, and am, pretty sure I did the right thing in Bolling's case. I talked to him man to man about a possible problem. He knew where I stood. I applied the golden rule, for I would have expected the same from him. Bolling confirmed, at least to my satisfaction, my belief that while seniors can't always be responsible for juniors, one can frequently judge the head by the body. As Arthur Newcomb once put it, "Show me the leader and I will know his men. Show me the men and I will know the leader."[4]

III

Circumstances were different in the case of Specialist Fourth Class Branch, who worked for me as a secret documents clerk. For a time, I was the only officer in the 343rd to have a top secret/crypto clearance. About all that meant, in practice, was that I was forbidden to go to Berlin, for fear that the communists would capture me, torture me, and extract from me (doubtless after a few seconds of resistance on my part) full accounts of America's nuclear secrets. (I wasn't at all sure of what I knew that could possibly be so important.) In any case, I became the unit's control officer for all our classified materials, which were numerous for our mission. Specialist Fourth Class Branch was an intelligent guy, and I had soon recruited him to be my clerk. He had the necessary secret clearance, and he responded well to my assurance that the work was "all indoors, with no heavy lifting."

The problem began when Branch submitted himself for a top secret clearance. All that was necessary for his job was the secret clearance he already had. But I think he regarded the TS clearance as a romantic thing, and he wanted it. Before long, an agent was examining me about Branch's background and interests, asking a long list of questions that were, I knew, necessary for Branch to obtain his TS approval. Then came the bombshell. I should have seen it coming, but I didn't. I now had to react, for I hadn't anticipated. "Do you have any reason to suspect that Branch is a homosexual?" asked the examiner.

The answer to the question was yes. I did have reason to suspect that Branch was a homosexual. But I had no knowledge of it. In Bolling's case, nobody ever asked me if I thought he might have used drugs. A problem never occurred. Branch's case was different. I had no certain knowledge that he was a homosexual. There were unkind allegations, from some people, however, that he was. An unfair and inaccurate stereotype exists of a homosexual. I refused to discriminate against him on the basis of innuendo. And he was a good clerk. But the question I was asked that day dealt with suspicion about his sexual orientation. It created a serious ethical problem for me. I flat out did now know what to do or to say.

If I responded no, that I was unaware of such suspicion, I would be lying, for Branch was considered by some to be a homosexual, and I knew that. But, on the other hand, if I answered yes, that I was aware of such suspicions, I could be slandering Branch, casting aspersions on him as well as calling into question my own judgment in picking him as secret documents clerk. Branch didn't intend to stay in the Army past his DEROS and outprocessing dates, but was it right, in effect, to label him a homosexual on the basis of rumor?

I told the truth—and perhaps slandered Branch. I tried to tell the investigator that I had no basis for my judgment. His question to me concerned an allegation. Clearly, in a court a question such as the one I had been asked could never have stood; here, however, it did. Suppose someone one night, in the EM Club, angered at a decision I had made referred to me as a "queer," and that Branch had overheard it. Would he have had to tell that investigator that he was aware of allegations about me, however ludicrous and unjustified? Probably. Suppose I had wanted to stay in the Army. Would my career have then been permanently damaged? As I recall, the question was moot in Branch's case, for his tour was up by the time the decision was made about his TS clearance. But it was brought home vividly to me that day why the courts have rules of evidence and established questioning procedures. And I have never felt right about my reply to that investigator, even though I can't be sure what alternative I had (other than challenging the entire procedure, which probably would not have helped Branch anyway). My understanding, by the way, is that the kind of question I was asked is no longer posed, although the

Army still discriminates against homosexuals. Whether it should or should not is not the issue here.

There is a kind of justice at work, however, even in the Army. Not long before my own DEROS, I discovered to my horror that I could not account for one of hundreds of secret documents I had responsibility for. Almost certainly it was an older, outdated document, probably destroyed as part of our normal culling process in which we tried, desperately, to minimize the number of documents we had. The problem was that for every document we destroyed (in a burn barrel—we didn't have shredders), we had to sign a CD (a certificate of destruction), requiring my signature and the counter-signature of a witnessing officer. Typing up the CDs with their elaborate and cryptic identification of the documents to be destroyed took hours. The destruction procedure itself took even longer, for we had to be perfectly satisfied that the CD matched the document, that signatures were affixed, that the burning left not so much as a trace of the documents still legible, and so on. In this labyrinthine procedure, one document that I believe we destroyed did not have the proper post-mortem CD. When we discovered that, I was again on the horns of a dilemma. If I reported the secret document as unaccounted for, as missing, it would mean a full-scale investigation with endless hours of questioning. Nobody wanted that.

On the other hand, if we prepared a CD indicating that the document had been destroyed, it was a boldfaced lie, requiring the counter-signature of another officer willing to participate in this "cabal." Dare I ask someone to participate? I could easily forge the signature of one of the officers who had signed CDs before. He would never know—among hundreds and hundreds of burned documents—that this one did not merit a real CD. Unless the missing document turned up, and the chances were a million to one against it, it is true that no one would have ever known. After a few years, even the CDs themselves could be destroyed. What to do?

I reported the document as missing. There was—there always is!—a regulation governing procedures in such matters, and I followed it punctiliously. It was a stupid thing to do. It was also the right thing to do. I was not naive, and I knew that, at the end of the road, there lay a black mark on my record as sure as the one that, however reluctantly, I had put on Branch's. When the investigation was done, I received a letter of reprimand from the battalion commander.

I was stung, deeply, by the letter. The investigator told me that the letter meant nothing. To me, of course, it shouldn't have, for I was about to DEROS; my tour of duty was almost concluded at the time. I had a number of laudatory letters in my file—none of them meant anything, either, since I was about to become a civilian again—and one perfunctory slap on the wrist was, I knew, meaningless. But I was crushed by the need, officially, to discipline me for my "malfeasance" as secret documents officer. Still, had I been the battalion commander, I would have responded in the same fashion. Nothing else occurred as a result of the unaccounted for document, although I will always wonder if it somehow took twenty years to blow its way into the Soviet Union, was finally discovered by Kremlin bureaucrats, and ended the Cold War. Probably not.

There is a footnote to this story. As I was leaving my battalion to go, where else, to Fort Dix to outprocess from the Army, I was talking to the personnel NCO. In the course of making small talk with him, I said that although my military record was undistinguished, I never expected to have a black mark on it. "What do you mean, sir?" he asked. I told him about the letter of reprimand. He told me that three copies of that letter existed. One would have been given to me. One was in battalion files, and it would be destroyed in a year. The last one—the only one that counted—was in my officer folder. "Is this it?" he asked. I replied that it was. "Very unfortunate," he said, "that it seems to have been lost [as he tore it into shreds and tossed it into his waste paper basket], and your official record has no mention of it."

"Thanks, sergeant," I said, willing to participate in this "cover-up." I have never numbered that among my most serious sins. And, by now, I really don't remember the name of that NCO. But thanks again, Sergeant.

IV

I got nailed as an officer for what, on balance, was a pretty minor thing—losing a document that I know we had burned. I could have been nailed for a lot of other things. There are times—as the personnel NCO knew—that the guy in charge can legitimately compromise and look the other way. Knowing when is critical. One night when I was the officer of the day, I had to inspect a group of MPs who were about to go on guard duty in our back area. At the 343rd we had "dog walkers." The dog walkers, or dog handlers (DHs—before baseball's designated hitters), were trained by the Air Force for sentry duty along with beautiful German shepherds. When their DH was rotated out of USAREUR, the dogs were destroyed. For a while the Army and Air Force tried to domesticate these killer dogs or to give them to new DHs, but it didn't work. So they were put to sleep when their DH left. I knew the DHs pretty well; I had to. One of the duties of the OD when he went into the back area was to agitate the dogs. Here is the procedure that we used.

The OD would take, say, an old hat and, while the DH strained at his dog's leash, the OD would smash the dog repeatedly in the face with the hat, putting the dog into a frenzy. The dogs were not muzzled, and I looked down the teeth of these wild-eyed, frenzied, powerful German shepherds many times. This procedure was necessary to maintain vigilant dogs. If any intruder got into the back area, he could have been ripped to shreds by these savage animals. Every time I agitated a dog, I did so with my left hand holding the hat. In my right hand, I held my service .45, loaded—not locked—without a round in the chamber (for safety's sake). That way, if the dog did get loose, I might have some chance of getting my pistol into action before the dog made mincemeat of my face and throat. One thing was certain: I tried to stay on the right side of those DHs.

The afternoon of that inspection, I happened across Specialist Fourth Class McCormick, who had been too long at the EM Club on post. McCormick was obviously drunk. "McCormick," I yelled, "don't you have duty tonight?"

"Yes, sir," he said.

"You know the rules about drinking during a duty day. I'm OD tonight, and I'd better not see you in [inspection] formation tonight."

He was immediately penitent. "Sir," he said, trying not to slur his words, "you can't do that! It'll mean I'll get busted!" He had that right. It would mean an Article 15 (of the Uniform Code of Military Justice) offense. He would indeed get demoted. "My wife's coming over, and I can't get busted! I can't make it [financially] as a PFC."

"Too damned bad, McCormick. You should have thought of that before drink number one today."

He persisted. "Please, sir, please! I'll be all right by duty time. I promise. Please."

Now what do I do? He had been in trouble before, and he knew all the rules. But he was telling the truth about his wife coming over. I knew he had been waiting for his promotion to E-4 before sending for her. If he was absent for duty because of violating the drinking rules, the matter would have to be referred to the C.O., who would almost certainly have to bust him if I turned him in. Should I look the other way? That was the easy thing and the popular thing to do. Should I call it square, as I had been taught to do? That was my duty. That was the tough call.

I made the easy call. Well, easy in a way. I spotted a couple of other DHs and called them over. "You guys are supposed to work together, right? Buddies and all that, right?"

"Yes, sir," they said in cheerful unison.

"Then prove it! You see McCormick here? He's drunk, and I have to inspect him and you guys at 1700. It's now a little after 1400. You have him in formation, as tight as a drum, at 1700 or I will have his ass. You understand me? You get him into a cold shower. You take him on a run. You pour an ocean of hot coffee into him. You make him upchuck from now until Tuesday. I don't care what you do, but you have him in that formation ready to pull duty at 1700 or he's as good as a PFC tomorrow!"

"Yes, sir." The language might have been hard, but they knew I was being soft. They knew I was giving McCormick a break but trying not to show it. I also knew they would try to have him ready for my inspection. Now I knew that the only thing that would do McCormick any good was time to get the alcohol out of his bloodstream. Showers and coffee would have no effect. But at least I got a team effort going. I hoped that, in less than three hours, McCormick's buddies could do something—at least have him in proper uniform. By not placing him on report, I was being derelict, and I knew it.

In less than three hours, we went through the guardmount inspection. I checked weapons and physical appearance and knowledge of guard orders—in theory. In practice, all I looked at was McCormick. There was so much blood in his eyes that he could have supplied the Red Cross for years. I had serious second thoughts. Had he been falling-down drunk, I would have had to put him on report. I thought he could do the job. As the troops were getting onto the two and a half-ton truck to carry them to the back area, I took one more shot at the happy warrior, whom I called aside. "McCormick," I said, "you understand something: you're not walking that dog tonight; I am. I don't mean I'm going to be out with your dog, but it is my butt on the line. I am giving you a very large break. You be damned sure you return the favor!"

"Yes, sir. Thank you, sir." was all he said.

When I checked the back area that night, I asked the NCO in charge of the guards how McCormick was. "He's O.K., sir," the sergeant said. I said no more. "Sir?" the NCO started.

"Yeah."

"You made a good choice."

"You just make sure that the men understand that I just used up all, every bit, of my patience tonight, sergeant."

"Yes, sir."

There were no incidents that night. McCormick's wife soon joined him and, to my knowledge, there never were any other problems with him. That time I was right.

Not too long after that, I was in the basement of the first platoon. An altercation had broken out between another young E-4 and an E-5 NCO. The E-4 had given the NCO some lip and, because I was there, I could not let the incident pass. I intervened, warning the E-4 that his reaction to the NCO could land him in a world of hurt, including the stockade. He was furious with the NCO and quickly transferred his anger to me, actually bumping me bodily, while contending that the argument was between the NCO and him, not me. There were a number of people who had seen this incident. I could respond by pushing or slugging Varner, the miscreant, or walking away with my tail between my legs, or taking administrative action. I figured if I hit him, I could hurt my hands—as well as lose face—or the fight! If I walked away, I'd lose any respect the men had for me. So the third choice was easy. "Varner, you get your act together right here and now! One more problem with you—in fact, one more word from you—and you're off to Mannheim [the major jail in the area]!"

Varner got the message fast. "Sir, I'm sorry. May I please talk to you for a minute?" I agreed, for he had quickly calmed down. The gist of his talk was that he thought the NCO he had argued with was being unfair to him, and he had "lost his cool." When I told him that losing his cool was not an option he had, and that neither was assaulting an officer, he seemed remorseful. I told him that I expected him to apologize to the NCO (and I would check on that), to get done exactly what he had been told to do, plus change his attitude in all he did. That was it.

Later an NCO asked me if I intended to file charges against Varner for assault. I told him that I was not planning to. "We were all hoping that you'd deck him," the NCO said. (I thought, "Thanks, Sergeant; Let's you and him fight.")

I said, however, "Things just got a little heated. Maybe I should have let Sergeant Smith handle it without intervening, but I didn't think I could ignore it. Maybe giving Varner a break will straighten him out."

The NCO was not impressed. "Varner's no good," he said, thinking that I had had a chance to convict him on grounds of assaulting an officer, and I had looked the other way. What really bothered me about this case was that Varner let it be known later that what affected my charity in this case was his witch-doctoring. As he told the tale, I had decided not to put him up on charges only after the doll he made of me, stuck with his pins, had unconsciously influenced me in that direction! Although Varner steered clear of me after that episode, he continued to be a pain to his NCOs, who certainly did not appreciate my willingness to pardon Varner that day. In this case, I was wrong.

V

I suppose that it is only God who knows perfectly how to temper justice with mercy. Sometimes I made the right calls; sometimes I didn't. That may be why, today, I am a little more forgiving of the umpires I draw as a baseball coach than are some of my fellow coaches. In this connection I made a call once as a first lieutenant that I am sure could have resulted in my hanging—or at least in my court-martial—had it turned out differently.

We had a barracks thief. Things were missing, here and there, from the wall lockers of guys in my platoon. We knew it was thievery, but the thief was not about to announce his identity to us. I called a leaders' meeting and told the squad leaders to remind their men that we had to work together, which ruled out thievery. Other than that, I left matters to my NCOs. Or I thought I had, until one day one of those NCOs came into my office to announce, "We caught him, sir."

"Say again."

"The thief, sir; we caught the thief."

"How do you know?"

"One of the guys caught him red handed. And we checked his wall locker, and a lot of the missing stuff was in there."

I was about to tell the NCO to get the thief over to the company orderly room where he could be read his rights and all the rest.

The NCO, however, had his own idea: vigilante justice. "We'll take care of it, sir," said Sergeant Gaines, in a menacing way.

"Sergeant Gaines, come here a minute. What, exactly, do you mean?"

"The thief is Tucker, sir, and poor old PFC Tucker is going to have himself an accident. He's going to fall downstairs."

Gaines was a pretty good man, and he was outraged by the stealing. We had been working on building up the spirit and community of our SASP platoon for many weeks, and Tucker had proved himself to be not only a self-made pariah, but a dishonest one at that, stealing from his fellow soldiers. Tucker had violated Gaines's and my sense of justice, and he was going to pay. If I had followed the book on this—that is, if I had done my duty exactly—my NCOs and my platoon would have been deeply upset. It was a "family" problem, and they wanted to solve it themselves. I wanted them to handle it. But I didn't want Tucker to suffer permanent damage as a result of our retributive justice, and I made my point clear to Sergeant Gaines.

I made a decision that OCS taught me not to make: to let my troops administer a beating to a barracks thief. But basic had taught me that there might be some value in this kind of action—when the guy who would not shower was forcibly showered, he became a model of personal hygiene. Which model to follow? If Tucker received permanent damage, we (Gaines and I) would receive permanent damage. And mine, rightly, would be the greater portion, for the decision and responsibility were mine. But if the beating did not take place here and now, under supervision, it might well take place later. Maybe Tucker would be better served, ironically, to pay up right then rather than later.

As it turned out, he did receive a noticeable beating, but the wounds were superficial and he healed rapidly. Moreover, in his own inarticulate way, Tucker made his apologies and was restored to a kind of guarded good grace in the platoon, a situation helped by the fact that he was overheard to tell other company members that he had lost his footing and fallen downstairs; whether his story was due to shame or fear, his version of events restored a little respect to Tucker.

So it turned out O.K.—a good decision? Wrong. From the standpoint of more than twenty years, I now believe I made the wrong judgment in this case. In letting a few people in the platoon have at this barracks thief, even though it in fact seemed to work out all right in the end, I put at risk Tucker, the NCOs, and myself. I bowed to the desire of my NCOs to teach the thief a lesson. Confronted with the situation again, I would have delivered Tucker to Army justice, not to the retribution of certain soldiers. Had Tucker been seriously hurt, my NCOs would have gone to jail; maybe I could have had a cell close by. Officers are supposed to be enough above the passions of the day to make reasonably disinterested decisions. I failed here. Although I carefully counseled Gaines not to inflict serious injury on Tucker—

proportionality was the idea—I had acted, in brief, as a trainee and not as an officer. I had chosen the wrong model.

To backtrack a bit to my first days at the 343rd, I recall a judgment that today seems open and shut that, at the time of its occurrence, did not seem too easy to make. Every young officer wants his NCOs to respect him, and mine seemed to be reaching that point after a few months of working with me. One night, as OD, I went to the NCO Club, that rundown excuse for a bar, and discovered a ruse in operation. The white NCOs had rigged the juke box so that the country and western tunes played right, but the soul music selections played country and western songs as well. The white NCOs gladly confided this to me when yet another black G.I. was robbed by the juke box that accepted his money and played songs he had not punched in. No refunds were forthcoming. The white NCOs were greatly amused at his exasperation.

"Terrific," I thought. "Now what? If I rat on these guys, I lose their respect and friendship. They have trusted me. But if I don't correct this injustice, I have been a white officer at the expense of black G.I.'s."

I saw this, after a minute of reflection, as a clear call. Respect on the wrong basis wasn't worth having. I went to the NCOIC of the club and ordered him to shut the juke box down at once; to get the wiring fixed before it went back into service; and to make refunds to any man, black or white, who had been cheated. A couple of the NCOs glared at me. If I had forfeited the respect of some of those NCOs, I at least still had my own. And I was still an Army officer—not a "white" officer.

A few months later, again serving as OD, I was called by one of the NCOs in the mess hall. "Sir, can you come over here? I got me an EM who won't follow orders."

"Now what?" I sighed to myself, as I began the slow stroll over to the mess hall. The NCOIC of the mess hall was irate. So was a PFC on KP that day. Both began to talk at once. "Hold it," I ordered. "Go ahead, Sergeant Winters," I said.

"It's a simple thing," Winters started. "Otter here won't clean up them pots and pans there, even though I ordered him to."

"Your turn, Otter," I said.

"Sir, it ain't that simple. Sergeant Winters promised me I could go after I cleaned up all the pots and pans, and I did. I got all cleaned up to go—I got a date, sir!—and then he calls me back, just as I'm leaving the mess hall and tells me to do these. That ain't right, sir. He told me I could go."

"Well, that's right, sir," said Winters, "but then we found this heap of pots and pans in another room. They gotta be cleaned, sir."

"But you told him he could go?" I asked.

"Yes, sir, but . . . ."

"Sir, I gotta get to my date. You ask Sergeant Winters, I did a real good job today."

"What do you think, Sergeant Winters, does Otter here have a case?

"Yes, sir, I guess. But I gotta get these things cleaned up, and it's his job to do it."

Otter's reply: "How come he let all the other KPs go, but he keeps me? Why is that, sir? Is that fair?"

"Why is that, Sergeant?"

"Well, sir, by the time I discovered these pots and pans, I had let everybody else go. I just barely caught Otter as he was leaving."

"Well, Otter," I said, "do you see Sergeant Winter's point here? He's got to get these things cleaned."

"Yes, sir, I do. But he made me a promise, sir. And I made my date plans based on that, sir, and I worked real hard."

They started to squabble again. "Knock it off," I ordered in a plangent voice. To myself I thought, "How do I get into these things?" Napoleon wonders whether to stay in Moscow. Lieutenant Toner wonders whether to stay in the mess hall. "Look," I announced, "Why don't you, Sergeant Winters, just help Otter here to do the remaining pots and pans? Good idea?" I thought it was an excellent idea! But Winters wasn't buying.

"Sir, I'm a *sergeant*. *Sergeants* don't do pots and pans."

"Yeah, well, they don't keep their word either, do they?" interjected Otter.

"Otter," I said, "will you kindly shut up for a minute." But the kid had a point. So did Winters. Well, if Solomon could risk dividing a child, I guess I could risk the honor of my lieutenant's bar. After all, I was an experienced pots & pans man from Fort Dix. In my best command voice, I said: "This is what we are going to do: Otter, start doing those pots and pans *now!* Sergeant Winters, you're helping him *now*. Before either of you say one more damned thing, I am going to help both of you *now*. Let's get this job done and shut up while we do it!"

I was fresh out of wisdom on this imbroglio. If either had fought that, I had nowhere else to go. *Mirabile dictu*, both seemed appeased, for neither had seen a lieutenant pull KP before. The job got done. A textbook answer to a weird contretemps? No. But you learn something very rapidly if you are perceptive as a young lieutenant: the "textbook" they talk about—it doesn't exist!

Field Marshall Bernard Montgomery once contended that "The beginning of leadership is a battle for the hearts and minds of men."[5] As concise and sententious as that maxim is, it overlooks something critically important. From the important problems to the silly ones, leaders, in peace or war, must be concerned with doing what they deeply believe is right; if they thereby earn the "hearts and minds of their men," as a byproduct, all is well. But if the leader's sole intention is to win respect, regardless of the merits or justice of his decision, he will win the hearts only of the deceitful and the minds only of the treacherous. Such is no prize.

VI

Among the lighter moments at the 343rd are the following, which will, in one way or another, seem familiar to all veterans. Almost everyone with some military service can recount stories similar to these. Perhaps the ultimate nature of the military—its responsibility to kill or to prepare to kill—causes servicemen to balance their terrible responsibilities with lighter moments. The West Point Cadet Prayer makes this entreaty: "O God, our Father, . . . [g]uard us against flippancy and irreverence in the sacred things of life." But that is not to say that, as *Reader's Digest* has suggested for years with its column about "Humor in Uniform," there cannot and should not be moments of some humor in the military. Fred Wenton knew that, and he was a better officer because of it. The sacred should never be confused with the profane; but at the same time, we should be spared from always taking ourselves too seriously. Maybe that is the real sense of the Cadet Prayer.

One day, the four-star general in charge of USAREUR was flying in for a visit: another dog-and-pony show! Now, when helicopters land, usually someone will "pop smoke," the green indicating permission to land and the pattern of smoke dispersal confirming wind direction and intensity for the helicopter pilot. On this particular day, the company clerk—no infantryman, to be sure—grabbed a "smoke grenade" and was about to pop it. An officer looked at the top of the grenade, which was color coded, only to determine that our venerable clerk was about to pop tear gas. I suppose that the helicopter would have landed O.K. to white "smoke," but I wonder how our four-star general would have liked walking into a storm of tear gas! By this time, my buddy Fred Wenton had gone home. Too bad—he would have thrown it!

Aside from attending Signal School while I was in USAREUR, I was also sent to a two-week Chemical/Biological/Radiological School (now called NBC for Nuclear/Biological/Chemical). Among the stories told there (and, from the accounts I have heard, it is true) is the tale of the lieutenant who was conducting CBR training for his unit. He was to use CN or CS tear gas grenades. Unfortunately, this lieutenant popped DM, which is a vomiting agent, not a tearing agent. DM is also used for crowd control, provided that your objective is to induce mass barfing, which does, I guess, prevent the crowd from doing much of anything else. DM, aside from its principal puking purpose, is odorless, colorless, and tasteless. The basic Army protective (or gas) mask affords complete protection against DM; but if you don't have a mask, you don't have anything to digest. CBR officers are supposed to have sense enough to check on wind patterns before doing CBR drills, but winds do shift. On this day the DM—which the protected soldiers thought was CN—was blown into a German town contiguous to the Army post. German men, women, children, cats, and dogs began throwing up lunch, breakfast, and last night's supper—for no apparent reason! Of course, the U.S. Army was soon implicated. I imagine that the poor lieutenant in this story is, by now, somewhere near the end of his very long incarceration.

The Army now pays most of its troops electronically. But in the Army of the early 1970s in Europe, officers still paid their own troops—all their troops whenever possible. That meant that we got to sign for thousands of dollars at our finance office (which became our money, kind of). Unfortunately, we had only a few hours to return to that finance office (1) all the money; (2) signed vouchers from the troops whom we had paid; or (3) some combination of (1) + (2) to equal 100%. If we lost money, we lost our money. If we "made" money, we were supposed to turn it over to Army finance. Funny how those things work. (It got even worse, by the way, when we had to serve as "D-mark" officers. One officer paid in dollars. Another officer was prepared to exchange the American dollars into German currency, employing the prevailing rate of exchange. I never knew one D-mark officer whose accounts were perfect at the end of a day of exchanging money for the troops. Some guys lost quite a bit of money—their own money. Some guys made quite a bit, by accident, of course—all of which was turned over to finance, of course, to help amortize the national debt. Of course.)

On one occasion when I was paying the troops, I had to make a side trip to the 97th General Hospital in Frankfurt, where one of our company's soldiers was. I wish he had been in the hospital because he needed something simple—say, open-heart surgery. But he was in the psycho ward. Does it make any sense to pay a man in the psycho ward? Whatever is he going to do with the money? But I had to pay him, then get him to sign a voucher that I had paid him. I had dealt with a number of psychos by this time, but this was the first time I ever had to pay one.

I left my driver and jeep at the door of the hospital, while I made my way to the psycho ward. I was going to pay PFC Wiley, and that's all there was to it. When I got to the ward, I explained my situation to the white-coated attendant, who knew the drill. I had to turn in my .45-caliber pistol, which was then locked in a safe. All the inmates wore regular fatigues. "It helps with their therapy if they think they're still on regular duty," was the explanation I received. Regular duty? When I went into the psychiatric ward, there were a couple of soldiers on regular duty who were urinating against the wall. All I wanted to do was find Wiley, pay him, get him to sign the voucher, and get out. There he was! "Hi, Wiley! How are you doin'?" (I'll admit that I was a little slow in the small talk department on this day.) To my delight, Wiley watched me count out the money to him. I had him count it back to me, as procedures dictated. He signed the voucher (with his real name—I checked). He then threw the money into the air! I figured that it was his money. "Good-by, Wiley."

I turned to leave and approached the door of the psycho ward. The shift of attendants had changed. I knocked on the door. "Let me out of here," I yelled. There was a small safety-glass window (the kind with wires running through it) on the door.

"Get away from that door," commanded the new attendant.

"I want to get out!"

"Everybody wants to get out. Now get away from that door!"

It didn't strike me as funny in the least at the time. Had I fallen asleep while watching some grade-B movie? How was I going to convince this attendant (1) that I was a lieutenant paying troops and (2) that I was sane? Better than that, that I was certified sane so that

I could work with nuclear weapons. (But would any really sane person do that?)

This contretemps did not last too long, thank God. "Hey," I yelled, "I'm a pay officer from the 343rd. How many guys in here have this on their collar [holding up the camouflaged bar of a first lieutenant on my right collar]?" That did it.

"Oh, sorry, sir."

Had he been playing with me? I don't know. I retrieved my pistol, clutched Wiley's voucher to my chest, and left. Happily.

I was sitting in my platoon office one day when I heard the report of a .45. I jumped up and raced outside. "What happened?" was my logical question. Two of my squad leaders—both of whom I had made sergeants (when they had been SP5)—had been planning to have a western-style shoot-out in the street. It was to be all kidding around, of course, with "empty" .45's. As one was drawing his .45 from his holster at "High Noon," it discharged, barely missing his foot (and, I daresay, other, perhaps more precious, parts, given its downward trajectory out of the holster). A .45 is fed with a clip or magazine (containing seven rounds) that is inserted into a well at the bottom of the pistol-grip. How in the name of heaven he could have thought the pistol was unloaded wholly escapes me. Nor do I understand how the weapon fired, for it has to be locked and loaded. (He must have cocked the piece as part of the show he was putting on.) He could have blown himself or his buddy away—all for the sake of a western shoot-out. I was not pleased.

Elements of these episodes are funny; elements are not, I realize that. The following account is not so much funny as it is peculiar, inducing a wry smile at best. At the 343rd, we had a great deal of racial tension. There were very serious flare-ups from time to time, and I do not mean to suggest that there was humor in their occurrence. But a sense of humor helped us deal with such problems wisely. I recall, for example, the gaucherie of one battalion commander I had. Determined to establish military order (not a bad goal), he decided to speak to the assembled company himself (bad mistake—he didn't know his own limits).

Despite the fact that this lieutenant colonel could never have been mistaken for a public speaker, he decided to address the entire company about discipline. That went O.K., given his marginal speaking abilities. Then he decided he would speak only to the black G.I.'s. I have no reason to presume he was prejudiced, and it does not reflect negatively on him that he had a deep southern drawl. Unfair though it might have been, many blacks regarded southern white officers with great suspicion. This battalion commander did not help our company's situation when he announced: "All right. All you white men outside. All you colored boys sit right here," which came out "rah-cheer." I closed my eyes and sighed. It wasn't the rah-cheer that bothered me. The black guys sat there glaring. I heard not a word of what the battalion commander was saying, although I guess it wasn't too bad. All the while I was thinking, "What do I do when these guys attack my battalion commander because he's an idiot?" He managed to leave post. His message was supposed to be that the Army cared for all its soldiers; but it came out that the Army had men—and boys. Today, I smile at that blunder—or was it a Freudian fumble?—of his. Then, all I knew is that I didn't want to be OD that night. There was a lesson in it: say what you mean. Mean what you say. And know your audience.

VII

I am truly persuaded that my time as a platoon leader has helped me greatly in my subsequent career as a teacher. The principal lesson I learned as a platoon leader, in one sentence, is that to be an ethical, efficient leader, one must be among the troops (students, parishioners, clients, patients, athletes, etc.) without being one of them. President Franklin Roosevelt said in 1940 that "A good leader can't get too far ahead of his followers."[6] The leader is the link between institutional imperatives and the followers' capabilities and desires. The leader can neither give into the relative complacency of the group nor establish goals so demanding that the followers cannot reasonably attain them. The leader owes great loyalty to his institution (the Army, a church, a school, a firm, a hospital, a team); but he also owes great loyalty to those who share his uniform or situation. That is the tension inherent in any leadership situation. In the end, the effective, ethical leader is well advised to learn De Pree's lesson:

> I would like to ask you to think about the concept of leadership in a certain way. Try to think about a leader, in the words of the gospel writer Luke, as "one who serves." Leadership is a concept of owing certain things to the institution. It is a way of thinking about stewardship as contrasted with ownership.[7]

NOTES

1. Russel B. Reynolds, *The Officer's Guide, 1967–1968 edition* (Harrisburg, PA: Stackpole Books, 1967), p. 34.

2. John Keegan, *The Mask of Command* (New York: Viking, 1987).

3. Max De Pree, *Leadership is an Art* (New York: Doubleday, 1989), p. 10.

4. Quoted in Robert A. Fitton, ed., *Leadership: Quotations from the Military Tradition* (Boulder, CO: Westview Press, 1990), p. 150.

5. Quoted in ibid, p. 150. See also S. L. A. Marshall, *The Officer as a Leader* (Harrisburg, PA: Stackpole Books, 1966), a revision of the government pamphlet *The Armed Forces Officer*, an excellent "leadership text."

6. Quoted in Fitton (above, n. 4), p. 150.

7. De Pree (above, n. 3), p. 10. See also Perry M. Smith, *Taking Charge* (Washington, DC: National Defense University, 1986), the introduction to which is by James B. Stockdale. Stockdale's works and that by James N. Rowe, *Five Years to Freedom* (New York: Ballantine, 1971) are excellent; both were POWs during the Vietnam War. Tragically, Colonel Rowe was murdered in the Philippines in 1989. His book is an enduring bequest to his countrymen.

5 CAPTAIN

> An officer's ultimate commanding loyalty at all times is to his country, and not to his service or to his superiors.
>
> —General George C. Marshall

> I have competed well; I have finished the race; I have kept the faith.
>
> —2 Tim. 4:7 (NAB)

An officer has twin responsibilities: to accomplish his mission and to look out for the welfare of his men. These are two principal ingredients of the military ethic, the topic of the final chapter of this book. But to talk about those two imperatives—mission and men—academically is one thing; it can be another to deal with the problems of men and mission every day, whether in war or peace. What follows is an account of mission and men in the 343rd Ordnance Company in the USAREUR of the early 1970s—an Army riddled with racism, drugs, and careerist soldiers intent only upon self-promotion. Of course there were professionals and professional temporaries. As there are always those who will burn the flag, so are there always those who will carry the colors. I think that men and women of substance seek challenge; men and women of substance then bring to the challenge they have chosen a commitment to do their continual, reasonable best. Whether they be in the armed services; in the church; in the academy; in the medical, legal, business, or athletic field, what principally matters to the professional is the quality of the effort, not the extent of the success. "[W]hile no man can be sure of the possibilities of his influence over other men," wrote S. L. A. Marshall, "every man knows by his own conscience when he is putting forth his best effort, and when he is slacking."[1] What follows is an examination of some soldiers whose efforts in the USAREUR of the day were by and large unnoticed by their countrymen but whose stories deserve a hearing. In all things, they did not succeed; but they tried to be true to the soldier ethic—to accomplish their assigned missions, however humble, and to care for their fellow soldiers during difficult and often bitter times. They put forth their best effort; and in so doing they fulfilled their duty. This is part of the untold story of a peacetime U.S. Army during the war in Vietnam.

I

Soldiers of the U.S. Army in the late 1960s or early 1970s dealt on a daily basis with racial tension. There was no escaping it. Every time a black G.I. saluted an officer or addressed him as "sir," he was in danger of being considered an "Uncle Tom" or worse by his black brothers, some of whom saw the Army as the spearhead of the military-industrial complex that oppressed black America. I did not share that view, and I was willing to talk at length to any G.I. in my platoon, black or white, about political and social issues. Actually, one of the few takers I had in that regard was a white E-4, a member of the pro-communist Chinese Progressive Labor Party. He was an informed and articulate advocate of Chinese communist ideology and of Marxism generally. Although I would hardly have had him as my secret documents clerk, that wasn't his job: he was helping out the unit re-enlistment NCO, covertly, I suspected, discouraging re-upping! In any case, the idea of rapping or talking things through was popular at the time and, up to a point, it had some merit. The E-4 I talked to was not anxious to read Bill Buckley, Pat Buchanan, or James Kilpatrick, but I think he was glad to have someone listen to his views carefully and debate them, I hope, fairly.

The Army has long discouraged, and for very sound reasons, doing what I was wont to do as a platoon leader. "Taking off the rank" in talking to the troops is not, my actions to the contrary notwithstanding, a wise idea. It can result in excessive familiarity, and it can disrupt the chain of command. But difficult times may call for different tactics. As I wrote in chapter 3, I had entered a whirlwind in the 343rd; I had to find stability in places different from those a platoon leader, in other times, in other places, might seek. As I have said elsewhere in these pages, there is no textbook that the beginning leader can read to find answers to all his problems. This does not mean that there are not enduring traits of a leader or that there are not perennial principles of leadership fundamental to virtually every circumstance; but how they are stressed or combined depends upon the man or woman on the spot. To lead is to improvise.[2]

There clearly was discrimination against blacks in the Army during these days. Frequently, the discrimination was subtle. The Army, after all, had pioneered the way for integration during the Korean War. But the policies decided upon at the level of the Department of the Army in Washington, D.C., surely didn't always find their way into the hearts of the officers and NCOs who were on the line daily, dealing with troops. And no dog-and-pony inspection would have turned up evidence of oppression against black troops. Ultimately unfair treatment of any race will end only when the human soul is converted; interracial harmony cannot be politically decreed from on high. (Such decrees, however, may well have a hortatory effect and are better, I suppose, than apathy in the face of wanton discrimination.)

The Army of my time had seriously begun to try to make things right for black soldiers. In housing affairs; in PX stocks; in commissaries; in chapels; in gyms, movies, and other recreational activities—in all these areas, the Army began to review its policies to determine if the interests of blacks were being served as well as those of whites. Perhaps the toughest job in the Army at this time was the job of the NCO, especially if he was a black man. Many very good black NCOs were dismissed by their young black soldier-activists as traitors to their race. The tragic death of Martin Luther King was fresh in the minds of many black soldiers, many of whom now believed that the path of violence was the only way to equality in a "racist America." I regarded that characterization of America as nonsense. But I had seen little of the oppression and racial hatred encountered by many of my soldiers. In some respects, theirs was a different world. I could do nothing to change their world or their past; but I could be sure that in my tiny part of the world, there would be no discrimination of any kind.

I meant that: of any kind. There was another kind of discrimination at that time and now—reverse discrimination. I knew some white officers and NCOs willing to ignore misconduct by blacks that they would have stopped or punished at once if committed by whites. I thought that was plain wrong. Some white officers declined to castigate a black soldier if he refused to salute. I regarded that attitude as wrong ethically and militarily. To the best of my knowledge, I never treated a black G.I. differently from a white one. I know that some would criticize that approach. But I am, at least, consistent. I believe that programs of affirmative action belittle and insult our black countrymen, who are—and who must be—able to compete on a basis equal with that of any other American. If whites or yellows patronize and condescend to blacks, African-Americans themselves, after a while, will no longer believe in the credentials awarded to them by well-meaning, if racist, whites who believe that blacks cannot compete, thus requiring that special accommodations be made. I believe that idea is pernicious, and that conviction sustained me in my tour of duty among black and white soldiers from all over the country. My line (and I guess it is corny, but I believed it) was that, in the U.S. Army, we were all green. I truly believe that the soldiers I led knew my views and that I was respected for them. I never had a problem with soldiers on the racial issue during my tour. Good men demand justice and fair treatment; they resent condescension.

Drugs were another matter. Despite my willingness to talk freely—and without attribution—to the kids I was responsible for, there was a lot of drug abuse in my platoon and company. I was in a very difficult spot because of the surety program. If, like a chaplain, I talked to a kid about his drug use, trying to help him or get him into a counseling program, I was supposed to report him as part of our surety program. Everyone in our company was supposed to be drug free. Even the officers (as in Lieutenant Bolling's case, I thought) were not entirely free of drugs—and certainly not free of alcohol, perhaps the worst drug of all. That could carry over to tobacco, too, if we wanted to talk about health. I was torn. I knew that my duty was to turn in, or report, soldiers who came to me during my PT (platoon time, or open-door time) and admitted their drug use. But I did not. Bolling, I thought, was a different case. I would have turned him in at once because he was an officer, and I held him (and myself) to a higher standard. (Would I hold majors and colonels to an even higher standard? Absolutely! The mistake I could forgive in a lieutenant I might well fire a colonel for.)

If I knew about a pattern of drug use over a long time, I was very careful, for unannounced reasons, about my assignment of troops. I made sure they were supervised and in non-sensitive jobs. At the time, there was no chance for troops to get drug counseling without running the risk of getting a less-than-honorable discharge. The problems were compounded at the 343rd where our isolation, our secret mission, and the times contributed mightily to our troops' boredom and sense of frustration.

I never betrayed a soldier's confidence, and what I was told in my open-door sessions remained confidential. I must say again here that I understand why the Army urges its troop leaders not to pursue this kind of policy; it can compromise a leader in the worst way. I was not a chaplain, and I was not a trained psychologist. Were I a company-grade leader today, I doubt very much that I would have the same policy. But this was not a volunteer army, with good discipline and high morale. It was a single step away from being an armed mob. I was in the midst of an epidemic and, although not a doctor, I was enough of a "medic" to apply some first aid and hold on. That is what we in USAREUR were doing in those days.

Friends of mine who led troops in combat suggested to me a leadership trick (bad word, really, for it's not intended to fool anyone) I often used. On sentry duty or patrol, leaders want their men sharp. The drug-crazy or spaced-out soldier endangers his buddies because he cannot do his job. The leader who says, "Get your ass in gear! You stay alert because I tell you to. And I'm the boss" is making an error, in my view. If he states it this way: "Get your ass in gear! You stay alert because I want you and your buddies to come back alive," he has it right. And he can say that (and should) in front of everyone. That is saying (1) we have a mission to do and (2) we will do it right *together*. He is also clearly indicating that he cares about his troops and will not "spend them" frivolously. Applying that lesson to garrison life in a peacetime Army is almost as important. The idea is to inculcate the spirit of the team.

I tried to do that by telling my people the obvious thing: if somebody doesn't pull his weight (if he's malingering or spaced-out, and so forth), we have to pull it for him. By eliminating Mickey Mouse stuff, by being fair and never playing favorites, by explaining (whenever possible) what we were doing and why, and by accepting (whenever possible) the suggestions of everyone involved, I think we began to create the teamwork that is needed in almost any military endeavor. Not for nothing does the Army want its officers to have some athletic background, for sports participation can teach that kind of lesson.

One of my charges, Private First Class Cann, was about to rotate back to the States for discharge. I knew he was on drugs. I knew that he had a very serious drug problem. Everyone knew it, in fact, and Cann would have been transferred out of the 343rd if his DEROS hadn't been imminent. I think Cann had fried his brain. He was barely responsive to me as he cleared post. He was like a robot. He appeared stupefied and torpid. He hadn't been like that months before. I tried, literally to the moment he left post to head back to the States for discharge, to get him orders keeping him in the Army on a medical hold so that he could get free drug counseling and attention from Army doctors increasingly (and necessarily) driven to deal with drug addicts. The day Cann left post to head home, on his way to freedom (as the G.I.'s called it), I am not sure that he even knew where he was going. The company driver had charge of him and would get him to the airport. Where he would go after the Army, I had no idea. I knew that my efforts to keep Cann in the Army for medical help had failed. In a sense, Cann was a casualty in a war just beginning.

We often failed in our drug-use deterrence programs. The sellers seemed always to be a step ahead of us. For example, some troops knew that heroin was pure poison, and they stayed away from it, preferring marijuana, which they thought, foolishly, to be benign. Some of the sellers laced their reefers or joints with heroin. Thus the soldier who wanted only pot often found himself desiring, for reasons he himself didn't know, to deal with only a particular dealer. Other dealers, to save supplies, sprinkled camel dung into the reefers. I am told, seriously, that camel dung gave the appearance and had the tactile qualities of pot, so the unsuspecting G.I. bought the reefer. Only later were his throat and lungs able to provide testimony about the bargain he had struck. Anyone who works, even for a short time, fighting drug dealers, as I did as an Army officer, begins to hope that there is a hell. For it will take an eternity of suffering to pay back some of these vile "human beings" for the misery and pain and suffering and grief they have inflicted on others through the sale of these poisons, these "recreational" drugs. The death penalty is too good for them.

Some of the troops knew that drug use could mean death. But the boredom, the lack of family and friends, the lack of a sense of purpose (were the Russians really going to attack?)—all these things contributed to drug use. NCOs I knew who had been in Vietnam said that the drug use in Europe among the troops was worse than in Vietnam. I am sure that it depended on the year and the unit, and I can't speculate about that. But I do know that the drug use was terrible in our company, supposedly a drug-free unit because of our mission! When the officers and most of the NCOs went home every night to their wives and their family suppers, they left behind young, sometimes frightened, and hopelessly bored young men, many of whom thought that no one cared about them. (That is one of the reasons I showed up in my barracks at odd hours of the night, just to be around a little, to "show the flag." Would I have loved, just once, to bump into my battalion commander in those barracks! But I never had as a commander a lieutenant colonel who did such a thing.) The troops thought our presence in Germany was a waste and, if their buddies in Vietnam were being excoriated (or, maybe worse, ignored) by the folks at home, how much worse for them? At least when a Vietvet came back, he had been to war. The USAREUR veterans could not claim even that, and they knew it. They were the truly forgotten. So they tried to forget, often by using drugs.

II

The upshot of all this, of course, was that the bad guys owned the barracks. I knew officers and NCOs who would not go into the barracks after 1700. It was too dangerous. I was always armed when I was OD, and there were times I had the staff duty NCO accompany me on my rounds. But when a United States Army officer is afraid of entering his own barracks, and insisting upon the observance of normal military courtesy, we had better disarm the Army at once! There were places I didn't *want* to go; but those were the places I *had* to go. I really never had a problem, after I was known. I insisted that rooms come to attention when I walked in, but then I would quickly put the men at ease and talk with them for a while. (As I have previously written, if I didn't respect my own rank, who else would?) My walks through the barracks weren't inspections. They were safety walk-throughs, and I never had problems, except for the occasional drunk whom I would order put to bed by his buddies. I would seize the moment to give a quick talk to them about watching him, lest he choke on his own vomit. But that would be the end of it. There are times and places to assert authority. And there are times to let things pass. Knowing the difference is sometimes tough; sometimes, it isn't.

When I was a new lieutenant at the 343rd, I was approached once by a group of black troops who, predictably, didn't salute. I stopped them. I had thought out carefully in advance exactly what I would do, if I could. This was it. This will not find its way into any Army text, but the gist of my "speech" to them was this: "Nobody here wants to be 'harassed,' right? Nobody wants NCOs or officers holding your hands and watching everything you do. I understand that, and you'll get nothing like harassment from me. You are bright enough and responsible enough to do your job. Do it. And nobody will play games with you. Saluting is a routine courtesy among people who wear the same uniform. I will exchange salutes with you when we meet. As a courtesy. That is the way it will be. We'll save a lot of time and trouble if we all understand this from here on out. Questions?" I got shrugs and sloppy salutes. Later, with the same group (whom I gradually got to know) I would cheerfully and strategically suggest that, "If you're doing a job—even saluting—do it right! Snap the salute! Show you're alive. Take a little pride!" Said with a smile, it helped; in my case, on a small post, it worked.

The same applied to haircuts, the bane of every officer and NCO at the time. I kept my hair very short, but I was willing to tolerate quite a bit from my troops, as long as their hair was clean and combed. There would come a time, however, when I'd have to say, "Smith, it's time!" and I'd nod toward his head. My line and logic were the same: "You don't want me to nag you on this, do you, Smith? Because I have better things to do! Do your job; look decent. No one will bother you. If you need an NCO or me to hold your hand, we'll do that, too—until we all get pushed into Article 15 stuff. And there's nothing to be gained by that." The appeal to the good side, to reason, to pride, to maturity—often said with a smile—usually worked. I know.

I know because once, getting out of my car in Frankfurt, I had my heels locked—that is, I was verbally assaulted—by a major because I was slow putting on my cover (hat) when I got out of the car. He thoroughly chewed me out in the parking lot of the PX. I had been chewed out at OCS by much more vitriolic guys than this clod. But I have not forgotten it in more than twenty years. I was burning! Even if I had committed some heinous offense (and I hadn't!), he could have made his point by saying to me, "Lieutenant, you can set a better example than that. I think you were a little slow putting your cover on and you should be more concerned about your appearance, O.K.?" He must have had a fight with his wife or something that day, so I became a target. I tried hard not to make a target out of my troops by abusing my authority with them. Most people will accept correction and criticism if they are offered in a gentlemanly fashion. Isn't that part of the commissioning ideal?

Take the case of First Lieutenant Carl Baines, a newly arrived officer at the 343rd. Baines was an ROTC officer who had made first lieutenant on the strength of a year's study for an MBA after he was commissioned. He was smart enough to ask for advice the first night he was OD, meaning that he ran the post in the absence of the CO. I went through the usual drill and then added some sage advice: "Carl, don't go to Ludwig's unless you have to." Ludwig's was a German-owned bar off post but in walking distance of it. I have talked about how I would not ignore a challenge and would insist upon proper soldierly conduct and bearing. But if I saw a group of troops walking toward our kaserne or post from Ludwig's and if I thought they were under the weather some, I would try to stay away from those troops. If the troops passed me without rendering the courtesies required, I would have to call them to task. That didn't help anybody, so I normally tried to avoid such meetings, especially if the troops weren't even aware of the encounter. If I had to pass by a wobbly troop, by the way, I would occasionally look the other way. There *are* times that insisting upon protocol makes one an ass.

That was my advice to Baines. The troops blew off steam at Ludwig's. The proprietors there generally managed things well. So there was no need for us to check it. I told him to stay away. But if a fight breaks out and if you're called from Ludwig's and if you must go there, go armed for bear! Take all the help you can get—the staff duty NCO, the CQ (charge of quarters), the company runner, the company driver, somebody's pet rat. Show force! Get the problem guy or guys out, and get out of there yourself. What I was trying to convey to Baines is that officers must try to maintain dignity. If an officer entered Ludwig's, some tipsy EM was bound to say something. Now what does the officer do? Ignore it? (Maybe. It depends.) Respond? (Maybe. How? It depends.) Best thing to do: stay away. Note that this was very different from staying out of the barracks, which were the property and sole responsibility of the Army.

Baines went to Ludwig's. He didn't have to. He just did. Somebody then criticized the President, or the Army, or all officers in history, or Baines's mother, or something, so he reacted. And overreacted. He ordered the troop outside. The troop was drunk, so he naturally told Baines where to get off, to the delight of his equally sloshed buddies. So Baines ordered them outside. Now you have an entire bar baiting this inexperienced, by now frightened, lieutenant. Baines got the names of most of the culprits and brought them up on charges. There should have been no need of that. Almost all of those troops were good soldiers who, although out of line that night, had no need of being placed "on the spot" by an inexperienced OD. Sometimes you call 'em right; sometimes you blow it. Baines blew it.

I had a leave coming at one point, and I filed the necessary papers. Rebecca and I wanted to go home to Massachusetts for a couple of weeks to let our parents see our son Chris, who was born in Frankfurt in 1970. My battalion commander kept me waiting for about three weeks, wondering whether he would approve my leave. I had to order plane tickets, and my time was about up. Finally, I called him. "Sir, Lieutenant Toner. Is my proposed leave O.K. with you, sir? I have to know because, if it is, I have to order tickets."

The response: "Don't bother me with that, Toner. I'll let you know when I've made up my mind."

Several hours later the colonel did call me back, approving the leave, and thus letting me purchase tickets—just in the nick of time—after causing me great exasperation and frustration. Why would anyone treat a subordinate like that? He confirmed something with me: I never kept anyone waiting who needed my permission for something! I couldn't always agree with proposals, of course, but I made a decision and let the person know, almost immediately. Some of this was common sense. If I intended to grant permission on something, why not do it with good grace, thus earning "interest" for myself? It sounds Machiavellian, I suppose, but why wait, and frustrate the petitioner, only to grant the request later, leaving a bitter taste in his mouth? I keep returning to the eminent good sense of the Golden Rule.

To conclude, I always found it very effective to ask troops to do things. I am not sure who thought up the absurd notion that officers never say "please" or "thank you," but it's nonsense. Of course there are times for point-blank orders. Period. But whenever I could attach a "please" or a "thank you," I certainly did. Timing is not everything, but it's close! There are times for confrontation and for screaming. But there are times, too, for calm and deliberation. Bishop Fulton J. Sheen once said that if you want to attract someone's attention, you should lower your voice; whisper, if you must. Too many NCOs and officers, on some kind of macho trip, want to prove their mettle and to exert their authority by yelling. Someone needs to take such "leaders" aside and say, "YOU FLAMING MORON! YOU SICKEN ME! STOP YOUR ENDLESS SCREAMING BECAUSE IT TURNS PEOPLE OFF!" And then say, "Can you see my point now? Are you listening more effectively now, when I talk with you? When I treat you reasonably? When each of us acts as a gentleman? Is either of us less of a man because we use our brains more and our vocal cords less? I want you to act maturely, so I treat you that way. Point made?"

Some problems, however, were of a much higher order of magnitude. The wife of an officer friend of mine, Mrs. Brown, taught at the post education center. I was at home one day when I received a call from a third officer who told me that "Roger Brown's wife was just raped on post. Roger's headed out to the post now. With a rifle. We've got to stop him." I raced to my car, intending to cut Roger off. I didn't make it. His wife was in the ambulance on the way to the hospital. After checking her and getting some information from her, Roger was ready to find her attacker and kill him on the spot. I found him as he was about to leave the education center. It was not a pleasant meeting. I told him how upset and angry and sorry I was, then pleaded with him to put down the rifle (shades of Staff Sergeant Colter!) and go with me to the hospital. "I'll go to the hospital and see Sue all right, but not until I kill the son a bitch who did this to her!" All Sue had been able to tell Roger was that a black man had been her assailant.

"Roger, for God's sake, what are you going to do? Start shooting at every black soldier on the post? You know better than that! Come on! Let CID [Criminal Investigation Division of the MPs] handle this!"

"Just shut up and get out of my way, Jim! I don't want to hurt you first—but I will if you don't move."

I didn't want him to hurt me either. But I *had* to stop him. He was crazy with anger, hurt and confusion. But I also knew that I would become the target of his hatred very quickly if I kept talking or didn't move. It was time to earn my pay.

Roger was a lot bigger and stronger than I was. He was armed. And he was enraged. I didn't doubt that he'd move me out of the way quickly, despite my hand-to-hand combat training (which I had never expected to use against a fellow officer). But I couldn't let him leave that building—not like that! I took a deep breath and prepared to tackle him when he made for the door. Maybe I could throw a lucky punch. Maybe.

Then the cavalry showed up in the form of Lieutenant Richardson. The balance of power and sanity were tipped in our direction, and Roger began to unwind a little. But he would not turn over his rifle, despite our entreaties. We did manage to get him off post, and we got him to the hospital. Mrs. Brown recovered physically, but it was a searing and terrifying experience for her. To be raped at the point of a knife on an Army post in broad daylight must still haunt her. The assailant was never found. Violence seemed always to lurk just beneath the surface during those troubled days. Breaking up fights and trying to head off racial episodes; drug counseling and maintaining good order and discipline—these were tough enough; but when you find yourself and a good friend about to fight it out as the consequence of a rape, it was altogether too much. And none of this yet describes the missions we went on.

III

As I have explained before, the 343rd Ordnance Company was a special (read: nuclear) weapons company, whose reason for being was to secure nuclear weapons and, if necessary, to distribute them to appropriate artillery and engineering companies. We also had a limited maintenance mission for those weapons. The site of my company was classified—rather ludicrously, for we all knew where we were! Although we were near such cities as Frankfurt, Darmstadt, and Babenhausen, we were isolated, exacerbating many of the troubles I have described. The exact nature of our mission is beyond the confines of this book, but some of our adventures can be described, surrealistic as they will seem.

In order to maintain the weapons we secured in our back area, we occasionally had to move them on or over German roads. The principal reason for the move was to get them to a special depot so that certain procedures could be accomplished. They would then be returned to the 343rd. Our special weapons warrant officers were able to determine when the moves were needed. As a SASP platoon leader, I then had to get the move done. That meant *convoy*. Now the word *convoy* means something very special to a Navy officer. It came to mean something very special, too, to this Army officer. I am part of a very small fraternity of officers to whom has fallen the responsibility of moving these weapons over German highways on which our good friends and allies, the Germans, place no speed limit. I thought it was bad enough on the highways. I was to discover it was worse in helicopters.

On a number of occasions, I served as the courier officer of a nuclear weapons convoy. When these weapons had to be moved, someone had to plot the move. That responsibility invariably fell to me in the 343rd. Several of us would gather around elaborate highway and topographic maps to plan out the best route to take. We would have to look for a wide variety of things, to include possible ambush sites by gangs who might want to steal the weapons. ("Good evening, ladies and gentlemen. This is Walter Cronkite. The U.S. Army, Europe, admitted today that twenty or more tactical nuclear weapons were seized this morning in a daring raid by a terrorist group known as. . . .") I knew enough to take the idea seriously. What a coup it would have been for some left- or right-wing nut group to steal nuclear weapons, perhaps taking the guards hostage in the process. So we always planned tight security for our convoys, drilling our troops repeatedly on the physical security measures to be implemented before, during, and after our convoys. The movements, by the way, were themselves classified secret or top secret. Army vehicles have their bumpers stenciled to indicate unit of origin. We had to cover the bumpers with burlap bags so that our company affiliation was thus, ostensibly, kept secret. Of course, the townspeople in the community adjacent to our post had known for years that we housed nuclear weapons; we just hoped that they wouldn't tell anybody.

A ground convoy consisted normally of a lead jeep; several large tractor-trailer trucks with the weapons banded, blocked, and braced in their beds; one or two two and one-half ton trucks with security teams; a communications vehicle (called a 26-Delta); a spare truck or two in case of a breakdown; and a trailing jeep. I was trained in explosives so that—really!—in case of an ambush, I could leave my jeep, climb onto the tractor-trailers (while my security team was in a life-or-death firefight with the gang intent upon stealing our weapons), rig up several shaped charge explosives, yell "fire in the hole," detonate the whole thing, and apologize to the gang because there'd be nothing left for them to steal. I wasn't sure I had watched enough John Wayne movies to know exactly how to get all that done, but that was my job. Most people figured that the explosion I was thus to set off would not go nuclear. We would just be denying the gang the fruit of their ambush. Of course, the whole scenario was incalculably stupid, but that was the drill. And we had to have the explosives with us on every trip. That particular regulation almost killed me—and my whole convoy.

After we mapped out our trip, we always did a reconnaissance to determine if the real path we had charted differed at all from the map's version. At least I got to use a little of my Infantry OCS training here. I must admit, frankly, that I really enjoyed this part of it. Planning and executing these moves was satisfying. We would get an order to move the weapons. Then, within the letter and spirit of the regulations and based upon a timetable we had to follow to some extent, the move was ours to make—meaning, effectively, my move. Every time I did a ground or air movement, I had a real feeling of accomplishment. It was great for the troops, too, because they were doing something; they got to be real soldiers for a change rather than showpieces during some inane inspection. On one convoy, Lieutenant Binder (the MP lieutenant who was to have so much trouble with Goppy Dolan) was my convoy commander. Although Binder and I had the same rank, he effectively worked for me on the trip. He was to provide security. As courier officer, I got to sign hand receipts for the nuclear weapons! (And I had paused about signing for D-marks as a pay officer!) They were "mine" until I delivered them and got somebody—the right somebody—to sign for them on the other end. You can bet that I was sure before I left post that *somebody* would be at the other end that day, patiently waiting for us to deliver our cargo. It was just like being a pay officer, with a much bigger payload.

Binder and I reconned the route the day before our mission. Everything was in order. We assembled early the next day. I did a final check—one of the rules of Rogers's Rangers was "Don't forget nothin'"—and we moved out onto the German highways. A convoy of this nature takes an incredible amount of planning and coordination. In a sense it is fun; but it is also in deadly earnest. We also knew that if a terrorist group hit us, they would be far better drilled and far better prepared than we were. We therefore had contingency plans in the event of an attack en route. The move was a caravan, but never cavalier.

I had to report in when we reached certain checkpoints along the route, all classified. A general officer, together with substantial infantry and emergency ordnance personnel (to include medics, MPs, and so forth), had to stand by until we got to our destination safely, and I had to maintain radio contact at regular intervals throughout the move. Army officers learn quickly the great truth of Murphy's Law, a small copy of which I used to keep in my wallet. But we had everything planned out on this one. We were on schedule, and everything seemed fine. I was in the trailing jeep, bringing up the rear. All the troops were in full battle dress, armed to the teeth. My jeep's trailer had the shaped charges, detonating cord, blasting caps, and various explosive implements I might need in case of ambush. We were to maintain a fairly high speed, considering the size of the vehicles we had, and not stop for anything except an emergency. We tried to set up MP roadblocks en route so we could breeze through intersections, something we coordinated by radio en route. It was sweet to see that work.

Suddenly something went very wrong. We came to a dead halt, and I couldn't understand why. There was no halt on the map, and Lieutenant Binder knew he wasn't supposed to stop unless.... Then I saw it. There had been no obstacles yesterday when we reconned the route; now there was road construction up ahead. We had to stop. Do I deploy the troops? Do I post perimeter security? I thought no, not yet, and I delayed giving the arm and whistle signals to effect the perimeter. On the back of my trailer was a large sign in German and in English, boldly proclaiming: "DANGER!!! EXPLOSIVES!!!" I may owe my life to that sign. Racing along down the highway came a German gasoline tanker, driving at customary—that is, excessive—speed. He managed to hit his brakes before plowing smack into the middle of our jeep trailer, laden with enough explosives to blow away most of our convoy and its secret contents. I was studying the map when, sitting in the shotgun seat of the jeep, I felt a terrific jolt and was pitched against the windshield, my steel pot bouncing off the windshield or the metal surrounding it. I was not hurt. An NCO in the backseat was sitting, gracelessly, between my jeep driver and me, knocked there by the force of the impact.

My worst fear had come true: it was a NAICP, a nuclear accident. Kind of. Should we put our nuclear accident and incident control plan into effect? I had this annoying habit of continually asking myself questions to which answers were not readily apparent. In a moment like this, one of the lessons I carried away from Fort Benning seized me. If you're in charge, be in charge, and act that way! The leadership reaction course at Benning had taught me that. The actions and orders came freely. I ignored the badly shaken gasoline tank driver. I called in a SITREP (situation report), while ordering security posted at once. Meanwhile I was trying to figure out what to do with the smashed jeep trailer. One of the NCOs suggested putting it in a reserve vehicle. Of course! Good idea! Our security team lifted it up onto the bed of one of our reserve vehicles, after removing those damned explosives. Then we got the hell out of there.

This was an accident that never happened. I could not tell any German police (1) where I was coming from or (2) where I was going. The movement was classified. I knew that, once my command had my SITREP, they'd respond appropriately. Lieutenant Binder, MP that he was, chalked the position of our jeep and trailer before we left. I never heard anything more about the accident; nor did I expect to. The German police doubtless ran into a stone wall when they investigated. Despite the stupidity of that tanker driver, he probably escaped blame for the accident; he was probably paid off by the Army. A Status-of-Forces Agreement allowed USAREUR to prevent German investigations into classified occurrences. This was very classified.

I had the consolation, as we drove away, that, had I been blown to bits, it would not have been a boring and routine death. How close did we come to annihilating a good portion of Germany that day? I don't know. But I do know that blasting caps can go off very easily. My ordnance people couldn't believe that they hadn't blown when our jeep trailer was crunched by that gasoline tanker. Could the shaped charges and other explosives have been blown, in turn, by the exploding blasting caps and by the other materials detonating in the trailer? Would the explosion have carried to the gasoline tanker itself? Would that, in turn, have led to such terrific heat that the nuclear weapons could have exploded as the whole convoy went up in flames? Would the nuclear weapon explosion have been only conventional—or would it have been a nuclear explosion? I am told that the possibility of a true nuclear explosion was really very small. My death would have been from a spectacular conventional detonation. Glad I missed it.

To this convoy, there is an ironic personal footnote. The day of my convoy, my wife Rebecca was to return to our home in Babenhausen from 97th General Hospital in Frankfurt with our newborn son, Chris. Understandably, she wanted me to pick her up at the hospital in our red VW beetle. I explained to her that I had a convoy that day and that she and Chris could take the Army bus home. It would deliver them to our doorstep—and it was free. I had a convoy, after all, and I had planned it. I wanted to execute it. It nearly executed me. Even today, I would feel justified in my decision to let my wife and newborn son come home on the Army bus if I could attribute it solely to duty. Professional soldiers—and professional temporary soldiers—must put *duty first* (that was, after all, the motto of my platoon). Were it to be otherwise, no soldier would ever leave his family. I am afraid, however, that all motives are mixed and, after more than twenty years, I think it's time for me to admit that, while I did go on that convoy from a sense of duty, I think the major reason I went on it was because I enjoyed it so much. This convoy, for many years, has been a source of dispute between my wife and me. She had always told me that another officer could have been the courier officer that day. My ego has led me to deny that for more than twenty years. This book forces me to be honest. I should have been at the hospital that day with my wife and son. I almost made it, but in a way I never would have predicted: I was very nearly at the morgue. If they could have found a few pieces of me.

IV

I didn't like Army bureaucrats hanging over my shoulder, even though they were useful from time to time. On another ground convoy, Lieutenant Binder's part of the convoy somehow got separated from my part. Somehow, traffic got between us and, at a fork in the road, Binder went left and my trucks went right. Terrific. There I sat on the side of the road, my security forces merrily motoring along to some unknown destination. A few drivers and I were left to protect tons of nuclear weapons with a few M-16's and a handful of .45's. Of course, I still had my trusty explosives—in a new jeep trailer. How was I going to call this one in?

"Wildcat six, this is wildcat niner. Over."

"This is wildcat six. Over."

"This is wildcat niner. We are having temporary mechanical difficulties. I say again: We are having temporary mechanical difficulties [such as half my convoy is missing!]. Over."

"This is wildcat six. Roger. How long will your delay be? Over."

"This is wildcat niner. [I haven't got a clue!] Anticipate no more than three zero minutes. Over."

"This is wildcat six. Roger. Advise if you encounter problems. Advise when you depart present location. Over."

"This is wildcat niner. Wilco. Out."

In about fifteen minutes, Binder and his lost legion made their way back to our location. Binder was grinning. Someone had told him about our "mechanical difficulties." "I should have reported 'meathead difficulties,' Binder. And you're the meathead!" I know. Shame on me! That is not the kind of honest report anybody needs to receive. It was kind of funny—but I was not being paid to be a stand-up comic. For this, I deserved a letter of reprimand. I had deliberately misled my command and, if anything had happened, I would have looked foolish and—what's more—untrustworthy. I still smile at this episode today, but I wouldn't counsel a similar approach to anybody today. I really had broken my own rule. I had taken the convoy in a cavalier manner. It worked out O.K., but my flip attitude wasn't called for. There are plenty of times to joke; but that wasn't one of them.

On the way back from this convoy, I must admit that I told another lie, although, I hope, a Platonic, noble lie. We had been having trouble with the troops shooting up atropine. In their protective or gas masks was an atropine syringe. Some troops contended that injecting atropine (which is an antidote to nerve gas) gave them a good high. In fact, it was extremely dangerous. Some of the NCOs were worried about the long drive back to post. We had unloaded our weapons, and all we had to do was to get there safely, but the troops had their masks (and their syringes). To take up the atropine would have been an insult to the troops; what was next—to take up their rifles? People do tend to act up to your expectations—or down to them. Before we left, I gave the troops a pep talk.

"You guys did one hell of a job today. We got the convoy done with only one minor error—half of you got lost! But in the end, we got the job done, and we did it right [well, almost right; I was overlooking my dissembling]. I'm proud of you! I'm also worried about you. Don't buy into this bullshit line about getting high on atropine. There are two things wrong with it. First, it can kill you—unless your buddy happens to hit you with a nerve agent right after you shoot up. Second, even if it doesn't kill you, it'll make you impotent. Now for those of you who almost joined the Marines instead of the Army, *impotent* means you'll never get it up again! So I suggest you think real hard about your girl friends. 'Cuz if you use this shit, thinking is the only thing that'll ever get hard for you again." It wasn't true; and that's not what *impotent* means. But nobody in our company ever used that stuff again. Did that make my little talk right? Once again, I wouldn't use the same approach today. Nor would my expression be identical to that I used then.

The troops enjoyed the helicopter convoys. And I did, too. Usually they went well—and quickly. The problem was that, during heli-convoys, as we termed them, I had to depend on some unknown warrant officers to be my convoy commanders. I had the security troops under my direct command, but the warrant officer-pilots (not to be confused with the ordnance warrants in the 343rd) were not always the best. On one particular mission, for example, we were flying east, toward the German Democratic Republic—East Germany. I familiarized myself with the means by which the pilots knew their location. I kept asking them, over the loud whine of the Chinook helicopter, "Where are we?" Usually, the chief would show me on the map. On this particular day, his reply sent chills down my spine.

"I don't know."

Terrific. (I got to say that a lot.) We were in flight toward East Germany with a cargo of nuclear weapons in a convoy of three helicopters and he didn't know where we were! Headline in the last edition of *Stars and Stripes*: "Lost lieutenant invades East Germany. Starts World War III." How about "Lost lieutenant Shot Down over East Germany with Chopper full of Nukes!"

"Find out where we are. NOW!" No *please* there! I was a little agitated. This time I had nothing even remotely witty to say. I whipped out my compass. If we were lost, I wanted to be lost flying toward France. Useless in the chopper. "Start flying *west* right now until you ascertain our position. I do not want to be *anywhere* near the East German border!"

Then he told me he couldn't get a reading.

I didn't want to try to navigate by the sun! So I reached into my pocket and pulled out a German Esso road map. If anyone has been in Germany, he'll verify how great those maps are. "Can you get down to treetop level? We'll use this." Even on an air convoy, we finally got our bearings by flying along reading road signs and following an Esso map. It worked. We finally got safely where we were supposed to go.

By the end of my tour, I was finally sure the best way to transport nuclear weapons: *don't!* I have to say, however, that the troops I had, given a serious mission and serious leadership (most of the time), responded effectively every time. I never had a single troop problem during any of the many convoys I directed during my tour. Cadets and candidates sometimes smile when they are taken through what the Army calls the "traits" and "principles" of leadership. That's too bad. I saw them work. And I saw a lot of American troops respond very well to field orders and missions that called upon them to do a serious job seriously. The only lapse in that regard was mine. They never let me down.

V

During any Army tour, soldiers have a number of personal encounters or experiences that deeply touch them and influence their decision to stay on active duty or to accept a discharge. I had never intended to remain on active duty, but one event that deeply affected me involved a brigadier general and one of my battalion commanders, the one who had separated the "men" from the "boys" during his homily about race relations. This particular lieutenant colonel (LTC) was not a dynamic speaker, but he seemed to be an honest and honorable man who, unlike another commander I was to have, would at least answer innocuous questions such as, "Sir, may I have a leave at such-and-such a time?" Lieutenant Colonel Sharp's boss, in our rather unusual chain of command (in what was called V Corps Support Command) was Brigadier General Walter Hummer.

Hummer reminded me of Goppy Dolan. Dolan's goal in life was to be promoted as fast as possible, and all of us who worked for him soon realized that we were minor weapons in the arsenal of his advancement. The same rationale motivated Hummer, whose immediate goal was the acquisition of his second star. That very obsession should have constituted ample reason for his not getting it! In any event, Hummer wanted all his battalions and their companies up to snuff, standin' tall, and lookin' good. Which was utter nonsense at the time. We didn't have the men, the material, the money, or the morale—let alone the training proficiency (given our turn-over rates)—to stand tall. Our job, of course, was to do the very best we could with what we had; crying about our situation wouldn't help. But the USAREUR of the early 1970s was, as I have pointed out at length, a disaster. After my tour, I knew that it would be only a matter of time until some long-range peace settled over Europe, for it was clear that the Soviets were not about to come through Fulda Gap. They must have known at the time about the advanced state of decay of the US Army, Europe. If they wanted to attack, especially with the war in Vietnam raging, they could hardly have chosen a more propitious time than in, say, 1971. That they didn't attack clearly indicated to me that they wouldn't—at least for a generation or two.

Still, our job was to prepare for that eventuality and to maintain the deterrence that, at least to some extent, engendered the end of the

Cold War. When his command had to submit REDCON (readiness condition) reports, Brigadier General Hummer did not want DA (the Department of the Army), meaning his bosses, to know the advanced state of decay of his command. So he was prepared to lie. Moreover, he expected his battalion commanders to lie. According to my sources on the battalion staff, Sharp refused to go along with the deception. His battalion was in no way or shape or form combat ready. We were *hors de* combat without ever having been in combat. And in the Army, looking bad is not just tantamount to being bad; it's worse. There were rumors about the grade or the ranking that Hummer told Sharp to report on his REDCON report. I do not know the details. I do know, albeit from others, that Sharp refused to lie on this official report. "To get along, you have to go along" is the wisdom. It reminds me of the school notion, used to condone cheating, that you "cooperate and graduate." Sharp refused to play the game. He was summarily relieved. The problem was not confined to Lieutenant Colonel Sharp alone. As two other Army officers of the period wrote, The U.S. Army Europe "was knowingly drawn down to support the war in Vietnam. Unfortunately, standards were not noticeably reduced as resources dried up—units with a handful of officers and rapidly rotating personnel were ostensibly evaluated by the same standards which applied to the pre-Vietnam 7th Army. The result was an intolerable stretching of the reporting honesty and integrity of the junior officer."[3] In Europe, the pressure on lower-ranking officers was to turn in high REDCON reports; that was bad enough. In Vietnam, the pressure on low-ranking officers was to turn in high body counts; that was worse.

Strictly speaking, it may well be the post hoc fallacy for me to contend that Sharp's peremptory relief was the product of his refusal to list his command as combat ready. Certainly, Hummer could not have written, "Lieutenant Colonel Sharp refuses to lie on the REDCON report, so I am relieving him of his command, effective immediately." But I had seen the raging epidemic of careerism before, with Dolan, and with many others. Careerism is the most visceral form of narcissism, for the careerist elevates himself and his private gain and personal interest above the health of his service and the honor of his profession. What he does in that regard is entirely at odds with the military ethic to be discussed in the final chapter of this book.

There were, of course, many happier moments. On 20 September 1970, I brought Rebecca to 97th General Hospital, awaiting the birth of our first child. I sat in the waiting room, in mufti (civilian clothing), worried and concerned as any father-to-be would be. Finally, the nurse appeared. She called my name, I thought, and I rushed over to her. "Congratulations! Your wife is fine, and you have a healthy baby girl!" I felt relieved and thrilled all at once. I asked if I could see my wife and daughter, and the nurse said I could, but just for a little while. When she directed me toward the room where my wife was, I thought she was mistaken. She was headed in the wrong direction. She, or I, was mistaken. "Excuse me, but you're not pointing me toward my wife's room." [I figured she must have flown helicopters before she became a nurse.] Then it dawned on her.

"Is your name Turner?"

"No," I replied. "It's Toner." When she said the name, I had nervously jumped up. Another guy—a Lieutenant Turner—was the proud father of the baby daughter he by now was probably thinking I had been trying to kidnap. When I jumped up, he sat down. Zero-sum game. Too bad he was so slow; had he been a little faster, he could have saved me some real embarrassment.

I sat down, feeling pretty silly about the whole thing—until I started worrying again. It wasn't too long until the nurse, very carefully enunciating the name, called for me and announced that my wife Rebecca and my *son* were doing fine.

In December 1971 my son Chris was about to have a sibling. I was hoping that Mrs. Turner wasn't about to deliver again, too. Rebecca went into labor, and I called the Babenhausen medics, who insisted on giving Rebecca a ride to the 97th General in the Army ambulance. (I had already assured Becky that she would not have to take a bus back this time.) I trailed along in our VW beetle. We knew it would be close; it looked as if Rebecca would give birth in the ambulance, much to the dismay of the young medic in attendance. I wondered how much training in birthing techniques he had received at Fort Sam Houston. My concern in that respect was interrupted by the horrifying sight of the back doors of the ambulance swinging wide open, as it charged along the icy roads to Frankfurt, revealing Becky on the cot with the nervous medic kneeling next to her. It was about midnight and the glare from my headlights and the ambulance's interior lights gave me a clear view into the ambulance. I expected Becky, our newborn, the medic, and a supply of medical goods to come tumbling out onto the road! Somehow the medic got the doors shut as the ambulance continued its race to the hospital. Never the world's coolest observer, I was pretty frantic about what I had seen—until it happened again! The doors banged wide open once again! This time the medic latched them. Good thing he didn't do that at first, or I wouldn't have had this story to tell! I got to spend the rest of the forty-five minute trip wondering if our child would be quicker into the world than the ambulance would be to Frankfurt. Rebecca never complained about that trip, explaining that calming the medic down throughout the ordeal seemed, in turn, to sedate her. Our son Patrick was born inside the Army Hospital in Frankfurt just a few minutes after Becky's wild trip there. There was no sign of the Turners.

Patrick very nearly didn't get born, not because he and his mother almost cascaded out of a careening ambulance, but because he very nearly didn't have parents. On a trip through the Swiss Alps not long after Chris's birth, we were headed down a hill—no! a mountain!—when I hit an ice patch and the VW began to skid to the side of the road. That, in itself, meant very little; but this side of the road was thousands of feet in the air, and there were no guardrails. I had not been driving fast, but the skid seized the car. There was very little I could do. I followed "the book": I steered in the direction of the skid and I pumped the brakes. I thought we were dead. Rebecca screamed, "My baby!" I kind of wanted to apologize for killing us all, but there wasn't time. One of my Monson High School classmates had died in a tragic fall off a Swiss mountain when he was in the Air Force, and I thought for sure we were following him. The car stopped. Barely.

I really hadn't been driving fast—which scared me all the more. We had a long way to go down that mountain. Rebecca started to cry. I felt terrible about our near calamity, and I was badly shaken, but I tried hard not to show it. Coming as it did not too long after I should have been blown up in the convoy accident, I began to look into life insurance.

We sometimes forget that lieutenants themselves are "kids." They can't be, of course, but they are. I really think it's helpful to senior military leaders when they recall their early days as an officer. I look back on some of these events today with amazement that we survived at all.

Chris tipped over a pot of scalding hot water on himself in Switzerland and by the time we were getting over the Pyrenees into Spain (I was really happy about being in the mountains again), we knew he needed to be seen by a doctor, so we managed to get to a U.S. Navy base in Barcelona, where Chris was treated for burns. He was O.K., and we continued our vacation trip—until I made a somber discovery. We were just about broke. By this time we were in southern France. We spent the last of our money filling up our tank and buying a few chocolate bars to sustain us. We couldn't stop anywhere, for we had no money to buy food or to rent a room. I figured that, if we drove all night, we could make the West German border in the morning. I had gas coupons good for the purchase of gas in West Germany; that would get us home. At about 5:30 the next morning, we crossed into the Federal Republic, and I pulled up to a gas station, having just made it. It didn't open for another hour, so we had to wait. In retrospect, I am glad I was able to transport, in the air and on the ground, nuclear weapons all over central Europe because I proved pretty incompetent at planning a simple motor vacation through Switzerland, France, and Spain. Aside from such inconveniences as no sleep, no food, no money, Chris's getting scalded, and our almost dying in the Swiss Alps, it was a fun vacation for a young couple.

My buddy Lieutenant Binder and I spent some time on the Kaiserstrasse, well known to many G.I.'s with time in Germany. The Kaiserstrasse featured some of the most elegant prostitutes in the world. It's not what you think. Binder was there as an MP officer, in mufti, but watching out for his men. I accompanied him, and our wives went with us, prudently going shopping while we patrolled the prostitute section of that famous street. It was a tough job, but .... Potential clients would discuss rates and, well, specialties with the ladies of the street, who rented booths to advertise their wares and to arrange terms with their customers. Deals made, the ladies would adjourn to rooms upstairs to bring the deal to a climax. The hallways were patrolled to prevent crimes against the ladies or their "johns." The ladies had their medical certificates posted on their booths, for they were routinely inspected. I do not want to advocate prostitution, but if it is going to exist—and it is—this was a pretty civilized way to go about it. As much of a cliché as it was, I actually did talk to a number of the ladies about "how a nice girl like you got into this business." Most of the prostitutes there, as opposed to the ones hanging around the post on payday, were educated women, many of them teachers, who were moonlighting to make extra money. Around every Army post, from time immemorial, have been these ladies of the evening. For the most part, these were civil and intelligent women, about whom, I hasten to add, I really know no more than that.

VI

It takes a special woman to be an Army wife. (I have no experience with Army husbands, whose wives are professional soldiers, but I can well imagine that a similar willingness to endure difficulties must mark them as well.) Rebecca was willing to stay in the Army. I was anxious to get on with my teaching career. Maybe she anticipated some of the deprivations we would encounter as I set out to earn my Ph.D. Some of the bad things at the 343rd I never told her; most I shared with her. In only one case can I recall her differing with me in strong terms. Good old Sergeant Meadows, the NCO I could not inveigle into helping us out in the back area when we needed to get ready for an inspection during his cherished time off, continued to plague the Army, the 343rd, and me. At one point, I had to do an EER (enlisted efficiency report) on him. I had no use for Meadows. Every month when he got paid he would take his money to the NCO Club, get roaring drunk, and stagger off post into the willing arms of the prostitutes. If Meadows was off duty, there was little I could do—as long as his conduct was not too flagrant. Somehow, Meadows walked the line. He generally, if reluctantly, did his duty; and he generally, and enthusiastically, spent his money on much wine and many women. If he wasn't discreet about it, neither was he pulling them onto the post. My hands were tied.

Meadows sickened me. He had an attractive wife, who was a very pleasant and thoughtful woman. He had five beautiful kids, who seemed to adore him. Yet month after month he came home with hardly anything with which to feed his family. His wife was reduced to using Army community (relief) services to get by. Officially, I knew nothing about this; but my wife (and others' wives) knew Mrs. Meadows. She had pride, and she didn't want others to know of her plight. But we did, even though there was little we could do for her.

Her husband's performance deteriorated. I have always believed that the mark of any man is whether, when he says something will be done, it will be. There is no excuse—short of death—for telling someone you'll do something and then failing to do it. If one is prevented from doing something promised, say by an accident, at least that commitment should be redeemed as soon as practicable, with notice given as soon as possible about the need for a delay. Such a trait is important in any man; it is critical in a leader. I couldn't trust Meadows. He would dissemble, deceive, and outright lie. He would say things were done when they weren't. He was unreliable. If our character is what we do, not what we say, Meadows failed badly on both counts. He hardly ever did what he said; he hardly ever said what he did. So on my EER, I told the truth.

Becky challenged me. "How can you do that to Lois?" she demanded. I tried to explain that I wasn't responsible for soldiers' wives. "You know that your EER will be responsible for getting him kicked out of the Army, don't you? How will Lois and her kids get by? At least here, in the Army, there are places she can go for help and still keep some pride."

"Do you think I should lie on the EER, telling everyone what a great NCO that bum is? Maybe I could get him promoted! Maybe he could become a platoon sergeant instead of a squad leader."

"That's not my business," was Becky's reply.

I won this one. Meadows was a fool and a liar. And I made the call on his EER. Then I told him, man to man. He was outraged and told me he was going to challenge it. I told him he had that right and should certainly exercise it if he thought I had been arbitrary and unfair. He began to complain about the Army, about life, about tooth decay. I didn't need to hear it. "You're dismissed, Sergeant."

Not long afterward, Meadows was transferred. If he "broke" my EER (which, in the 343rd, also required an endorsement from my boss, the company commander, who agreed with me), I never found out. But the need to distinguish between Rebecca's humanitarian view and my professional view has stayed with me. Even today, as a professor, I must distinguish between, say, a student's need to graduate and the grade I assign him on the basis of what he earns. There is a place for mercy. There is a place for justice. Only God always knows which is which. The rest of us must do the best we can, trying not to be too strict when some mercy is called for or too lenient when the truth demands an exact accounting. There is always the chance, too, for forgiveness, for everyone makes mistakes, some of us serious ones. In Meadows's case, I suspect that both God and his wife were ready to forget and forgive, depending upon his contrition and amended ways. But I had to call the EER square and did.

VII

In October 1971, I was promoted to Captain. Although I was an infantry officer, I had never worked a day in my MOS. All my active duty time, to date, was spent as a nuclear weapons officer. It had taken me only twenty-six months to pass from second lieutenant to captain. In fact, I was in the first wave of officers for whom twenty-*six* months were necessary for those promotions. Before that only two calendar years had been required because of the demands of the Vietnam War. I had just gotten myself promoted out of my platoon leader's slot. I became installation coordinator for the 343rd, meaning that I supervised plumbing and electrical work.

I knew that I would soon have to make a choice about staying in the Army. A reduction in force (RIF) would swing into high gear before long, and volunteers to get out of the Army were being called for. Our battalion S-1 (personnel) began a poll of the battalion's officers: Who wants to get out? I raised my hand. My battalion commander asked me to consider remaining in the Army.

"Thank you, sir, but no. I just want to teach."

"You can teach in the Army, Toner."

"Thank you, sir. I'll consider it."

I did, and I didn't.

I had been one of the few officers at the 343rd not relieved. It had been that bad. I had seen a bevy of captains (and battalion commanders) come and go in the chaos and turmoil of USAREUR in those days. I had gone through all the emotions, I think, that any soldier goes through—except for the horror and sorrow of losing men in combat, an experience I am deeply grateful never to have had. But I had lost men to accidents, to drugs, to racism, to ignorance, to poverty, to illiteracy. I had had a number of successes and a number of failures. I was proud of my achievements and ashamed of my defeats. At OCS, we had had it drilled into us that the officer sets the example: "Follow me." I had done my utmost to be a competent, organized, fair, and enthusiastic officer; to get my jobs done efficiently and reliably; to look out for my men and to show compassion. I had made a lot of calls. Although I had made mistakes and misjudgments, they were not the mistakes of a physical or moral coward, and they were not the misjudgments of the unprepared or the apathetic.

Shortly after I received my orders telling me that in sixty days I would DEROS and be honorably discharged from the service, I went to the Officers' Club in Babenhausen. There I saw an artillery captain literally crying in his beer. He had received the same notification I had, and he would soon be getting out of the Army. I had asked for it; he hadn't. He had never finished his college degree; his Artillery OCS diploma wasn't enough to keep him safe from the RIF. His career was done, well short ahead of retirement. I managed not to disturb him with my pleasure that my Army career was over, too.

Just before I left the 343rd, the NCOs had a surprise farewell party for me, and they presented me with a music-playing stein. The song was the lovely *Auf Wiedersehen*. This was the first time such a party had been given there by the NCOs, and I was deeply moved. I think my eyes misted over as I tried to be humorous while trying to convey, in a short talk, my real sentiment: thanks. I treasure that stein. During the party, the company first sergeant, an old E-8 about to retire himself, said, "You know, you're all right, Captain." Because I had never heard him say that about, let alone to, any other officer, I keep the remark framed even today. Right in the center of my mind.

On 2 March 1972, I was separated from the Army, with Rebecca, Chris, and Pat in tow. When we got home to Monson, Massachusetts, I took off my greens for what I thought would be the last time. (Never did I think I would end up teaching in a green uniform at Norwich University.) I had to stay in the inactive reserve a little while longer, but I had so much active duty time I didn't have to go to any meetings or drills. In May 1974, I received my honorable discharge as Captain, Infantry from the United States Army. It was my birthday.

VIII

There were thousands of us, soldiers like me, who had joined the Army for a cluster of reasons, some pretty solid, some pretty suspect. We were never sent to Vietnam. We would have gone. We were soldiers. We were never sent, though many of us volunteered to go. And then we were civilians again. We tried hard to do our duty, most of us, as we had been given the light to discern that duty. It was odd, somehow, to be, not a Vietnam veteran, but a Vietnam-era veteran. There seemed to be more passion, and more pain, in serving in Vietnam or in fleeing to Canada. The story of the Vietnam-era veterans is not told in this account. I am not every Vietnam-era veteran. Most of us Vietnam-era veterans feel that the monument in D.C. is ours, too, for all of us knew men who had served there, some of whom had fallen. But perhaps this account tells something of the joys and sorrows of the officers and men whose youth was spent, not in the battles and backplaces of Vietnam, but in the confusion that was the U.S. Army, Europe, in the early 1970s.

Most soldiers, after all, are not heroes. And very few are villains. If they are honest, they serve. That is enough. They spend their time, their talent, their hopes, their dreams, their fears, and their sorrows in the service of their country for two or three or four years. Some give their all; some give a lot; some give a little; some give nothing. They are human beings, with all the sins and sufferings of humanity written about through the ages. But in the end, the Vietnam-era veteran, whether draftee or regular, wore the uniform of our country in the conviction that it was an honorable thing to do. There are ever the knave and the rascal; there are ever the knight and the gentleman, and the uniform rarely changes the fundamental character of its wearer. The Vietnam-era veterans were—and are—of all stripes and accounts. But they are inextricably bound by the cords of having worn the uniform at a critical time in their nation's history. And they are linked to their brothers and sisters of Vietnam combat service in the knowledge that we, too, answered the trumpet. We, too, carried the colors. We, too, served the Republic. We may not share their Vietnam experience, but we share some of their sorrow, some of their pain, some of their pride. And we thus come to terms with them, with ourselves, and with our country, which we served with true faith and allegiance.

NOTES

1. S. L. A. Marshall, *The Officer as Leader* (Harrisburg, PA: Stackpole Books, 1966), p. 239. This is an excellent book for officers of all grades; it appeared earlier as *The Armed Forces Officer* (Washington, DC: Government Printing Office, 1960), published as DOD Pam 1-20. Although Marshall was its author, he is not credited with authorship because it is a government publication. One of the finest books on leadership I know is David Donovan, *Once a Warrior King* (New York: McGraw-Hill, 1985). "Donovan" is a pseudonym for Dr. Terry Turner of the University of Virginia, an adviser to the Vietnamese during the war in Vietnam. This book, subtitled "Memories of an Officer in Vietnam," is an extraordinary account. The best general treatment I know is that of Sir John W. Hackett, *The Profession of Arms* (New York: Macmillan, 1983; this is reprint of the original volume published in 1963).

2. Three books, in particular, might be recommended to the young company grade officer. Although two are referred to earlier in the notes to preceding chapters, they are well worth another mention. Edward M. Flanagan, *Before the Battle: A Commonsense Guide to Leadership and Management* (Novato, CA: Presidio Press, 1985); Dandridge Malone, *Small Unit Leadership: A Commonsense Approach* (Novato, CA: Presidio Press, 1983); and Aubrey S. Newman, *What Are Generals Made of?* (Novato, CA: Presidio Press, 1987).

Examples of other books that can be consulted with profit by the company-grade officer (aside from those already referred to, such as *Platoon Leader* and *Company Commander*) are these:

Lee G. Bolman and Terrence E. Deal, *Reframing Organizations: Artistry, Choice, and Leadership* (San Francisco, CA: Jossey-Bass, 1991), which along with many other management-hint books are worthy of perusal; James H. Buck and Lawrence J. Korb, eds., *Military Leadership* (Beverly Hills, CA: Sage Publications, 1981); Robert A. Fitton, ed., *Leadership: Quotations from the Military Tradition* (Boulder, CO: Westview Press, 1990), an interesting collection of apothegms; John Keegan, *The Mask of Command* (New York: Viking, 1987), or any of Keegan's other volumes; Lloyd J. Matthews and Dale E. Brown, eds., *The Challenge of*

Military Leadership (Washington, DC: Pergamon-Brassey's, 1989); Robert L. Taylor and William Rosenbach, *Military Leadership: In Pursuit of Excellence* (Boulder, CO: Westview Press, 1984); Martin van Creveld, *Command in War* (Cambridge, MA: Harvard University Press, 1985).

Reading histories and biographies is essential to developing a sense of leadership. If one reads with an eye to "picking up tips"—that is, to learn lessons applicable to one's own life and situation—that reading can be invaluable. But the key is to read widely and well, as well as to keep organized notes (capable of being retrieved according to one's own system) on those readings and one's reflections on them. Elsewhere in this book I strongly recommend Roger H. Nye, *The Challenge of Command: Reading for Military Excellence* (Wayne, NJ: Avery Pub. Group, 1986); it is so well done that it's well worth another plug.

3. Zeb B. Bradford, Jr. and Frederic J. Brown, *The United States Army in Transition* (Beverly Hills, CA: Sage, 1973), pp. 237–238.

6 FOUNDATIONS OF PROFESSIONALISM: LIBERAL EDUCATION AND ROTC

In God's name, I beg of you to think.[1]

—Demosthenes to the Athenians

To create the power of competence without creating a corresponding sense of moral direction to guide the use of that power is bad education.[2]

—former Dartmouth President John Sloane Dickey

When I removed my uniform that day in March, 1972—and received my honorable discharge in May 1974—I felt a curious ambivalence. On the one hand, I was glad to get out of uniform and return to civilian life. On the other hand, I respected the uniform and what it stood for, and I had been proud to wear it. When Rebecca, my two kids, and I returned to my home town of Monson, Massachusetts, I was faced with (1) unemployment and (2) the need to feed my family. As a veteran, I was entitled to unemployment compensation for a time. One day, when applying for it, I was confronted with a situation that gave me a very fleeting and minor insight into the humiliation that can attach itself to such things.

Now unemployment compensation is not welfare. In a sense, I had paid for this kind of insurance protection while I was in the service. And I truly was seeking a job at the time. I punctiliously addressed the clerk filling out my form as Mrs. Smith—only to have her condescendingly refer to me as James. I resented that. I was a college graduate, a former Army officer, the father of two children—a responsible man—and I did not like being patronized by a clerk, whom I was respectfully addressing. This single episode confirmed in my view the importance of getting on with my life—fast.

When I was a senior at Saint Anselm's, I had applied to a number of graduate and law schools, among them The College of William and Mary. On my way to OCS at Fort Benning, I had passed a sign in Virginia pointing the way to the college. I had regarded it wistfully then, but now I was intent on continuing my education at that institution, the second oldest (behind Harvard) in our country. Another veteran and one with whom I was to become quite friendly at William & Mary, was Joe Seymour. When Joe first applied to W&M, he had sent an envelope (complete with return address), but had absent-mindedly neglected to enclose anything with it! He was too embarrassed to follow up. A few days later, he received a "reply" from the Chairman of the William & Mary Government Department: "Was there anything you wanted to tell us?" With red face, Joe applied, finally, and was accepted. Both of us had had our plans for a master's degree interrupted—mine by design, his by the draft—by the Army. We pitched in, now, in 1972, with frenzy.

As I was finishing my degree at W&M, I began the process of applying to doctoral programs. My choice was clear: Notre Dame. When the University offered me an excellent opportunity, I was determined to accept the challenge. Thus in August 1973, Rebecca, Chris, Pat, and I piled into our old red VW beetle to begin the trek to South Bend, Indiana. Once again I threw myself into my studies with complete effort, something permitted only through the understanding and help of my generous wife, Notre Dame, and the G.I. bill!

I became friendly with a group of Air Force officers who were at Notre Dame pursuing their Master's degrees. One of those officers, Vic Willwright, helped us move into an old house on South Bend's Eckman Street after we had lived for several months in cramped university apartments. Another officer, Tom Moore, was a gregarious guy and a serious student. I used to kid him all the time about being a "lifer"—something I could get away with because all the active duty Air Force guys knew I had been in the service. One day another Air Force officer at Notre Dame and I were talking. He casually mentioned that Moore had spent five years at the "Hanoi Hilton" as a POW. I was stunned beyond words. Tears came to my eyes when I reflected on how I had been lampooning him about his time in the Air Force, his "lifer" mentality, and the rest.

Later that day, I found Tom and sought to apologize for my insensitive and stupid remarks. "You haven't said anything to offend me, Jim," he said. "I figure you're one of us." That off-the-cuff remark, coming from him, meant and means a great deal to me. In classes, Tom could hear remarks, from students and professors alike, about "baby bayoneters," "mad bombers," and so on, and he would respond with equanimity. If, however, someone alluded to Jane Fonda, he would excuse himself from the class, returning only when the subject had changed. "This is a university," he told me, "and if I can't discuss a subject rationally, I should not participate. And I can't be rational about Fonda." Fonda had coolly proclaimed that Americans were being treated well by the North Vietnamese. While she was there, advancing that absurd thesis, Tom Moore and his brothers in captivity were being beaten and abused by their captors. But Moore refused to be baited, even unintentionally. I have always thought it nonsensical that the silly movie *An Officer and a Gentleman* concerned a fictional hero in officer training. At Notre Dame, I had met yet another real officer and gentleman. Years later, in my first lecture to the Air War College Class of 1991, I made a silent "dedication" of my remarks to Tom Moore.

After finishing work on my doctoral dissertation, I had a couple of months before graduation ceremonies, so I requested permission to teach in Indiana schools to supplement my income. At Saint Anselm's, I had practice-taught at Manchester (New Hampshire) Memorial High School and, after I left active Army service, I had taught again for a little while at Monson High School in my hometown. I knew I wanted to teach—that was all I ever wanted to do, and it was one of the attractions of being an officer. Little did I know that everything was about to come together in a curious way.

There were very few jobs available in 1976 for nascent college teachers. But teaching again in northern Indiana schools confirmed my desire—my *need*—to teach. A position opened at Notre Dame, and I competed for it. I was offered a one-year contract teaching in a great books seminar, supplemented by teaching one course in International Relations. That job was a life-saver, for it fed us while I sought a tenure-track position elsewhere. I thought that, because of my background in the military and my academic interest in world politics, I might be able to compete for a position at one of the country's military colleges, such as The Citadel or Virginia Military

Institute. I did get a response, of sorts, from Dr. Kenneth Russell, of Norwich University, in Northfield, Vermont. Norwich did have an opening at about the time I would be finishing up my one year of post-doctoral teaching at Notre Dame.

As it turned out, I was invited to Vermont for an interview. I was in heated competition with another would-be Norwich faculty member who had just earned his Ph.D. at the University of Virginia. The Norwich faculty split down the middle about the appointment, half for me and half for my rival. It was Dr. Russell's call, and he chose me. Thus in late fall of 1976, I knew that the next summer would find us back in New England. Life is filled with ironies. First, I would be back in the uniform of an Army officer, for the Norwich faculty—and those at VMI and at The Citadel—teach in uniform, although they are civilians. Second, my competitor for the Norwich job is, as I write this, on the faculty with me at the Air War College in Montgomery, Alabama. He has been very content in central Alabama. I have been content in central Vermont. How different our lives would have been had Dr. Russell decided differently in his tie-breaker vote!

For more than a decade now, I have been teaching at a unique institution, Norwich University. Founded at Norwich, Vermont in 1819 as the "American Literary, Scientific, and Military Academy" by Captain Alden Partridge (a graduate and former Superintendent of the U.S. Military Academy at West Point), Norwich was the first institution to offer an organized curriculum in civil engineering. Norwich was the first private military college in the United States, and its guiding philosophy led to the passage of acts establishing agricultural and mechanical arts and to the Reserve Officer Training Corps. (Justin S. Morrill [1810–1898] was a Vermonter representing his state in the House of Representatives [1855–1867] and in the Senate [1867–1898]; it was the Morrill Act of 1862 that established land-grant colleges and universities.) Indeed, the current command historian of the U.S. Army Cadet Command is clear that "the concept of the Reserve Officers' Training Corps (ROTC) . . . [was] put . . . into practice with the founding of Norwich University (NU) in 1819."[3]

To understand Norwich and its problems is to understand the genesis of ROTC in our country and, at once, to understand the awesome complexity of the professional military ethic in a democratic setting. Norwich is the home of the "citizen-soldier," the man (now the person) trained generally in the military arts and sciences who, like Cincinnatus when his country is imperiled, rushes once more to the colors and to his country's defenses. It was, after all, Woodrow Wilson who argued that "We must depend in every time of national peril, in the future as in the past, not upon a standing army, nor yet a reserve army, but upon a citizenry trained and accustomed to arms."[4] Wilson's views—and those of Captain Alden Partridge—seemed confirmed again in 1991 as the United States drew upon its Guard and Reserve units to provide strength and depth for its commitment in the Middle East.

In the fall of 1977, I returned to uniform as a faculty member at Norwich University, the spiritual father of ROTC. With some trepidation, I embarked upon my teaching career at an institution called a "military college," whose cadets and faculty, although civilians, wore uniforms; whose curriculum mandated instruction in and by ROTC instructors; whose students were organized into a corps of cadets; whose central idea was that of imparting a military education. Was that an oxymoron? Was a military college a contradiction in terms? Could there be real academic freedom at a university whose premier purpose was to produce Partridge's citizen-soldiers?

I

As the nation's only private military college, Norwich is unique. Virginia Military Institute, in Lexington, Virginia (founded 1839) is, in the judgment of one historian of the military colleges, "*the* surviving classic military college"[5]; The Citadel, in Charleston, South Carolina (founded 1842), and North Georgia College, in Dahlonega, Georgia (founded 1873), are all publicly supported institutions. These three colleges, and Norwich, are "essential military colleges," meaning that they comply with standards established by the Defense Department to merit that label. Texas A&M University, in College Station, Texas (founded 1876) and Virginia Polytechnic Institute and State University, in Blacksburg, Virginia (founded 1872) also have corps of cadets but are not regarded as essential military colleges, for they do not comply, in all particulars, with the stipulations demanded for that certification. There are also a number of military junior colleges and institutes (such as Georgia Military College, Kemper Military School and College, Marion Military Institute, New Mexico Military Institute, Valley Forge Military Junior College, and Wentworth Military Academy and Junior College). These institutions, complemented by a number of state maritime institutions and the federal academies (such as West Point) constitute the military colleges in the United States. (The next chapter will examine the armed service War Colleges, such as the Air War College.)

In addition to the four military colleges (whose freshmen, by the way, undergo military training as rats [VMI], knobs [The Citadel], rooks [Norwich], and frogs [NGC]), there were in the 1988–1989 school year 315 Army ROTC units (scheduled to be reduced soon by fifty or more), 66 Navy ROTC units, and 151 Air Force ROTC units. There are also a number of high school units. In 1916 the National Defense Act created the Reserve Officers' Training Corps to establish a pool of trained officers. Navy ROTC was set up in 1926; Air Force ROTC, in 1946. ROTC is the primary source of officers for all of the armed forces, but most colleges (e.g., William & Mary and Notre Dame) offer ROTC without having a uniformed corps of cadets as one finds at the four military colleges (and at Texas A&M and at VPI). In brief, Captain Alden Partridge and Norwich University began quite a program with the creation of the small Vermont college in 1819!

As I joined the Norwich faculty in 1977, I was acutely aware of the traditions of ROTC, although I had not been an ROTC-commissioned officer myself. I was aware, too, of the persistent danger of military education: vacuousness and intellectual imprisonment. Robert E. Lee himself once said that "The great mistake of my life was taking a military education."[6] And on the south gate of Norwich University was inscribed this shibboleth: "Liberty is obedience to the law." I had seen that motto just once before—on a sign over the stockade at Fort Dix, New Jersey!

My purpose here is not to examine the educational regimen at the federal service academies. That has been examined, with varying degrees of objectivity, before; and it will be again.[7] There is comparatively little work about the curricula at the military colleges, and that is regrettable because of their unusual status. Would there be too little serious scholarship at VMI and the others? Would they be too dedicated to the martial arts or to athletics to be serious about learning? I wondered if these schools had anything of substance to offer the nation. George C. Marshall, for one, had thought so. "This institution," he said of VMI, his alma mater, "gave me not only a standard for my daily conduct among men, but it endowed me with a military heritage of honor and self-sacrifice."[8]

West Point and Annapolis aside, the education of ROTC officers is critical, and ROTC has never enjoyed high academic prestige.[9] One thing was certain: if the military colleges, the flagships of general ROTC instruction in the United States, could not offer serious and substantial academic education, the chances were that ROTC as offered in regular colleges would be vacuous. Jefferson's dictum—"We must train and classify the whole of our male citizens, and make military instruction a regular part of collegiate education. We can never be safe until this is done"[10]—would be absurd unless the military colleges were, first and foremost, *colleges*. Military education would indeed be a contradiction in terms unless the leaders of the military colleges, and their ROTC instructor counterparts on the civilian campuses—could comprehend that what matters is *education*; the military part of it referred only to the fashion in which the education was offered at VMI, The Citadel, Norwich, and North Georgia or, at most colleges, the occasional wearing of the uniform during ROTC drill.

It always surprises my colleagues at other institutions that, in Norwich's political science department, we do not award academic credit for ROTC. Indeed, I have long been strongly opposed to that pernicious idea, for ROTC instructors are not hired or promoted or tenured in response to obtaining academic standards and procedures. The ROTC curriculum continues to be weak and unchallenging and serious students have traditionally found very little of intellectual interest in that curriculum. ROTC is vocational instruction; it is doctrinaire instruction; it is intended to serve a particular ideological purpose. That is all right, provided it be understood that, of necessity, ROTC is tendentious—that is, ROTC instruction is not intended to be an objective, open-minded survey of a subject. It is intended to produce an ensign or second lieutenant. It should never be confused with serious-minded academic instruction. In the end, ROTC is training; it is not education—an idea that must be examined here before we explore professional military education in the next chapter and the idea of the military ethic beyond.

II

The manner in which, of late, education has been expected to toe the socially relevant line is wholly at odds with any genuine understanding of what liberal education is all about. The renowned *Oxford*

English Dictionary provides this definition of *education*: "culture or development of powers, formation of character, as contrasted with the imparting of mere knowledge or skill." The ancient Greeks had a name for such enterprise: *paedeutics* (also *paideutics*), which meant the science or art of education, and the development of *paideia* in students. In addition to having a name for their effort at educating, they also had a model of both an educator and of a truly educated man. In the *Apology*, Socrates remarks that "[L]ife without enquiry is not worth living."[11] That, coupled with Plato's magnificent allegory of the cave (in Book Seven of the *Republic*), should be the paradigm for liberal education. Briefly put, this idea of liberal education means that we must recognize our own benightedness and limitations and attempt, throughout our lives, to turn to the sun of real knowledge, away from the shadows of foolish opinion. Thus, the truly educated person is a paradox: he knows that he does not know, but he recognizes the enormous value of wondering and of questioning. Such a person is marked by a genial skepticism, but not by a radical nihilism. That the educated person will have the true strength of humility does not mean that such a person must deny all dogma, principle, or faith. While the educated person will not claim infallibility or omniscience, he or she must develop sufficient power of discrimination to separate the wheat from the chaff, the permanent from the ephemeral, the meritorious from the meretricious. Of course, such moral and mental discernment is not developed in the saccharine courses of many of today's "progressive" schools. Nor is it developed by curricula that offer, say, recreation or hotel administration as a university major.

If all of this sounds academic and pedantic, I would defend it by contending that, since 1946, and the end to the defense of superior orders (*Respondeat Superior*) at the war crimes tribunals, soldiers must be able to distinguish legal from illegal orders and ethical situations from unethical—no mean task! An education that does not empower the soldier—especially the officer—to make such judgments is not only a failure; it is a danger. An education conducted in the absence of some awareness of, or appeal to, some standard of agreement that men and women of ordinary sense and common decency can identify is, according to the Greeks and Jews and Christians of antiquity, hardly worthy of the name.

John Henry Cardinal Newman (1801–1890) contended that "Liberal Education, viewed in itself, is simply the cultivation of the intellect, as such, and its object is nothing more or less than intellectual excellence."[12] Newman would have agreed (at least in one respect) with McGeorge Bundy, who wrote in *Daedalus* that

> The university is for learning—not for politics, not for growing up, not even for virtue, except as these things cut in and out of learning, and except also as they are necessary elements of all good human activity. The university is for learning as an airplane is for flying.[13]

Although liberal education is sometimes imbued with mystical qualities or is associated with religious sectarianism, a Catholic prelate—again Cardinal Newman—expressly disavowed the notion that liberal education and seminary education are necessarily coincidental:

> Liberal education makes not the Christian, not the Catholic, but the gentleman. It is well to be a gentleman. It is well to have a cultivated intellect, a delicate taste, a candid, equitable, dispassionate mind, a noble and courteous bearing in the conduct of life;—these are the connatural qualities of a large knowledge; they are the objects of a university.[14]

We return again to the idea of an officer—and a gentleman. The officer who is not a gentleman will prove himself the scoundrel, unworthy of the confidence reposed in him by virtue of his commission. Thus, to the officer corps, the quality of the undergraduate experience is critical.

The modern, progressive curriculum—saturated as it is with vocationalism and vacuousness—has sunk to apologizing for the liberal arts as "communications skills." (Norwich, for example, followed a trend in establishing a major in communications when it already had an English major. Norwich also offers a *major* in recreation, as well as in physical education—an oxymoron, if one ever existed! VMI, to its great credit, offers no majors in such things.) The Newmanesque notion that the university's premier objective is the cultivation of the intellect; Bundy's hortatory idea that the university is a center of learning rather than a meeting hall for dissidents or a hang-out for post-pubescents; and the corollary idea that a substantial liberal arts education ought to develop in students a gentility, a refinement, and a bearing in keeping with the spirit of their introduction to the life of the mind—these are hardly popular ideas today. On too many campuses, the liberal arts have become the *soft* alternative to the supposedly *hard* engineering and science courses. The purpose of the liberal arts—like the *raison d'etre* of the university itself—is obscured.

III

As Newman pointed out, the liberal arts should not be parochial or sectarian. Still, study of the liberal arts should lead one to understand that not all values are equally meritorious. As Russell Kirk put it, college "is meant to join knowledge with virtue."[15] To Kirk, good knowledge is wisdom, and the ends of education are wisdom and virtue. "Wisdom," he says, "means apprehension of enduring reality; virtue means the development of strong moral principles and habits."[16] Thus, the study of the liberal arts should help students to think clearly, to write concisely, and to speak cogently. But—what is far more important—the cultivation of a discerning, critical intellect and the development of a taste for serious reading and reflection ought to help send students from the university "into the world with a cast of character fitted for ethical and intellectual leadership."[17]

Colleges worthy of the name, although open to the great questions (as C.S. Lewis put it in *The Four Loves*, "Whatever is not eternal is eternally out of date"), do not mandate answers. In the past, particularly on military college campuses, the danger has been from the "right." Beginning in the mid-1960s and continuing today, a quarter-century later, the danger on most campuses is from the "left." On the campuses of today, for example, one is expected to be politically correct. As John Leo has put it:

> Student bodies have always tended toward generational groupthink, but the new orthodoxy is unusual. Its purity is guarded by faculty who rebelled as students in the 1960s. Affirmative action, busing, gay rights, women's studies, the PLO, animal rights, bilingualism, the self-segregation of blacks on campus and censorship in the pursuit of tolerance are all politically correct. The following are non-P.C.: The SAT, doubts about abortion, Catholics, wearing fur, any emphasis on standards or excellence, and any suggestion that gender and ethnicity might not be the most overwhelmingly important issues of the modern era.[18]

As in the 1960s, today's ROTC students need to have a kind of mental toughness, first, to find a curriculum that will be serious and challenging and, second, to resist the "PC" ideological "purity" one finds on many campuses.

IV

Of course, one can question the psychological make-up of anyone enrolling in ROTC. Are such students ill? Are they fascists? Are they ethical prostitutes, selling themselves out for their monthly stipend ($100.00, by the way) or their tuition scholarship? At least one writer has suggested about 1960s ROTC students that, compared with those who were draft dodgers or resisters:

> [The ROTC students] . . . were found to have experienced strict childhood discipline in relation to a prominent father figure. They showed a strong concern about proving their masculinity, used more alcohol, felt powerless to influence their country's actions, felt troubled about their sexual inadequacy, defined independence as loss of self-control, preferred a well-ordered and structured environment, admitted being self-centered and egotistical, felt shy with girls but boasted to their fellows of their sexual conquests, claimed little real intimacy and poor relationships with the opposite sex, admitted treating females as objects, tended to seek dominance-submission relationships, and were relatively aggressive, impulsive, irresponsible, and non-intellectual, with a poorly developed conscience.[19]

On balance, this is a pretty sick group of people. Aside from the fact that one wonders if the study (based on a small sample of cadets at the University of California) was accurate then—how would the far-left student movement leaders of that day have stacked up?—one can ask whether it would be on target today. Winners of the four-year ROTC scholarship, for example, are clearly among the best students in the nation. West Point students (and their peers at the other federal service academies) are traditionally excellent students, good athletes, and proven student leaders.

The point of all this is that, although one can dismiss ROTC students as psychological misfits (more discussion of this idea can be found in the next chapter), in fact such criticism is little more than an absurd *ad hominem* type of argument. If ROTC itself continues to survive as the primary source of officer procurement in the United States, perhaps it is important to question the fundamental character of those who enroll in the program. But one needs little familiarity with psychological research to dismiss such tendentious studies as the one previously quoted as of little or no value to serious inquiries into the ROTC program.

V

If ROTC students are not always psychiatric basketcases, they are students seeking an education. If we regard education as an enterprise prudently and at least partially concerned with imparting sound character as well as knowledge, genuine liberal education concerns far more than business education or management training. As Walter Lippmann once put it:

> If it is the role of reason to be an instrument of each man's career, then the mission of the schools is to turn out efficient careerists. They must teach the know-how of success and this—seasoned with the social amenities and some civic and patriotic exhortation—is the subject matter of education. The student elects those subjects which will presumably equip him for success in his career. The rest are superfluous. There is no such thing as a general order of knowledge and a public philosophy, which he needs to possess.[20]

Lippmann's principal point in *The Public Philosophy* is, of course, that such a general order of knowledge does exist and that the college curriculum that fails to impress upon its students the vitality of such a moral hierarchy is derelict. The renowned naval strategist, Alfred Thayer Mahan, wrote, even before the turn of the century: "So I think any scheme of education is defective that makes no effort to teach the learner to believe in and depend upon God."[21]

Whether one prefers reference to "divine law," to "natural law,"[22] or to "moral law," the liberal arts education must concern ethical conduct. Indeed, the inchoate military officer must—to repeat, *must*—have some familiarity with the codes and canons of ethical conduct before he is commissioned. Such familiarity cannot be derived from the sciences alone. The student must have a grounding in literature, philosophy, political science, history, and theology. Yet ROTC scholarship students may not major in theology, and majoring in philosophy is discouraged.[23] As we will see in the next chapter, when they enter senior service schools, professional officers are expected to know and then to develop their ethical knowledge (and, presumably, conduct). The late, renowned scholar of international relations, Hans Morgenthau, wrote that

> The moral law is not made for the convenience of man[;] rather it is an indispensable precondition for his civilized existence. It is one of the great paradoxes of civilized existence that—in contrast to the existence of the animals and barbarians—it is not self-contained but requires for its fulfillment transcendent orientations. The moral law provides one of them. That is to say, human existence, not in its animal but in its civilized qualities, cannot find its meaning within itself but must receive it from a transcendent source.[24]

Although Morgenthau's religious heritage differed from that of Newman, he very probably would have agreed with Newman that "truth is the object of Knowledge of whatever kind."[25] Today, even the idea of "truth" is ridiculed by a pervasive nihilism that laughs at the idea of duty or responsibility or obligation.

> What Moses brought down from Mt. Sinai were not the Ten Suggestions. They are commandments. *Are*, not *were*. The sheer brilliance of the Ten Commandments is that they codify in a handful of words acceptable human behavior, not just for then or now, but for all time. Language evolves. Power shifts from one nation to another. Messages are transmitted with the speed of light. Man erases one frontier after another. And yet we and our behavior and the commandments governing that behavior remain the same.[26]

One wonders what kind of preacher or right-winger may have spoken those reactionary sentences. It was Ted Koppel, at Duke University, in 1987. His argument *must* be understood, at least by those empowered to determine the ROTC curriculum! A first lieutenant in the field is told by his company commander, a captain, to "waste" a group of unarmed civilians, including women and children. Is the order legal? Is it ethical? To what standard can the lieutenant appeal if he should choose to disobey it? Unless that lieutenant has been educated in the school of "truth"—that is, in the conviction that a natural or moral law exists that transcends the positive, practical law represented by the captain—we will have another My Lai massacre on our hands, the product, to some extent, of a poorly educated lieutenant with no frame of reference to employ as he sorts through orders of dubious ethical content.

At the Air War College, I teach a course to colonels and lieutenant colonels that I call Command and Conscience. By the time I teach that course, these officers have made scores of command decisions, many of which are, at bottom, ethical questions. If their education in ROTC, OCS/OTS, or at an academy has been lacking in the firm guidance that study of the liberal arts should impart, they may have already committed a series of ethical blunders of significant magnitude. None of this, by the way, is intended to say that studying ethics—especially military ethics—or being a major in the liberal arts is a guarantee of right conduct. But it *is* to assert that studying such subjects at least holds out the prospect that the student has, presumably, thought about some tough issues; and that is better than nothing!

VI

Russell Kirk has strongly argued that "a principal achievement of liberal education in America has been the imparting of a sense of moral worth among the more intelligent of the rising generation."[27] This sense of moral worth is often called *conscience*; as a former president of Dartmouth, John Sloane Dickey, put it thirty-five years ago, "the American liberal arts college (including the church colleges) can find a significant, even unique, mission in the duality of its historic purpose: to see men made whole in *both* competence and conscience."[28] This recurring term *conscience* is integrally connected with the liberal arts endeavor, the recent politically correct nonsense notwithstanding. As explained by Kirk:

> Really "conscience" means ethical judgment in the light of certain enduring ethical principles, tempered by the cardinal virtue of prudence. Unless one recognizes and understands norms, unless one's conscience is tutored by authority, what is mis-called "conscience" becomes self-righteous prejudice, glorying in one's own rectitude.[29]

A similar judgment was made by a Navy admiral, H. G. Rickover, who contended that

> Morals are the quarrel we have with behavior. Yet, any system of education which does not inculcate moral values simply furnishes the intellectual equipment whereby men and women can better satisfy their pride, greed, and lust.[30]

Precisely so. One often hears it said that a liberal arts education is "liberating." In one sense, of course, that is correct. The person liberally educated should be able to escape the chains of rank prejudice, gross ignorance, and perpetual slavery to his or her own basest instincts and urges. But, in a different sense, a genuine liberal education—to be vigorously differentiated from what passes for one in some vacuous curricula or in some politically "correct" course—is a captivating experience. Development of informed conscience also requires commitment to high purpose. The educated person loses himself in devotion to ideals that, at least in some substantial part, transcend his own interest.[31] Like life itself, liberal education is paradoxical.

There is, finally, no better explication of the ideals of education itself than this brilliant passage from Cardinal Newman, well worth quoting at length:

> It is the education which gives a man a clear conscious view of his own opinions and judgments, a truth in developing them, an eloquence in expressing them, and a force in urging them. It teaches him to see things as they are, to go right to the point, to disentangle a skein of thought, to detect what is sophistical, and to discard what is irrelevant. It prepares him to fill any post with credit and to master any subject with facility. It shows him how to accommodate himself to others, how to throw himself into their state of mind, how to bring before them his own, how to influence them, how to come to an understanding with them, how to bear with them. He is at home in any society, he has common ground with every class; he knows when to speak and when to be silent; he is able to converse, he is able to listen; he can ask a question pertinently, and gain a lesson seasonably, when he has nothing to impart himself; he is ever ready, yet never in the way; he is a pleasant companion, and a comrade you can depend upon; he knows when to be serious and when to trifle, and he has a sure tact which enables him to trifle with gracefulness and to be serious with effect. He has the repose of a mind which lives in itself, while it lives in the world, and which has resources for its happiness at home when it cannot go abroad. He has a gift which serves him in public, and supports him in retirement, without which good fortune is but vulgar, and with which failure and disappointment have a charm. The art which tends to make a man all this, is in the object which it pursues as useful as the art of health, though it is less susceptible of method, and less tangible, less certain, less complete in its result.[32]

The way to think about this, in a military connection, is to ask whether, if such a description could be accurately made of a new college graduate, as he received his degree, one would feel safe in commissioning him as a second lieutenant that same day. Kipling expressed it this way:

> If you can talk with crowds and keep your virtue,
> Or walk with Kings—nor lose the common touch,
>
> If neither foes nor loving friends can hurt you,
> If all men count with you, but none too much;
>
> If you can fill the unforgiving minute
> With sixty seconds' worth of distance run,
>
> Yours is the Earth and everything that's in it,
> And—which is more—you'll be a Man, my son!

Cicero, who died in 43 B.C., thought that genuine education was "a spur to the young, a delight to the old; an ornament in prosperity, a consoling refuge in adversity."[33]

VII

But what, finally, has all this to do with the way of the warrior? What has all this to do with blooding the enemy's shirt? With finding, fixing, fighting, and finishing him? Everything. "An army of deer led by a lion," Chabrias (410–357 B.C.) is supposed to have said, "is more to be feared than an army of lions led by a deer." Remove the educated, purposeful, analytical officer from the army and you remove its central nervous system. It was John Milton who said that "I call, therefore, a complete and generous education, that which fits a man to perform justly, skilfully [sic], and magnanimously all the offices, both private and public, of peace and war."[34] That is—or was—exactly the purpose of Norwich University, and of its founder Alden Partridge, and of the ROTC of which Norwich is the progenitor.

But many technocrats and progressives might object that Newman, Kipling, Cicero, and Milton have nothing to say to us today. (Let them read Koppel's remarks again!) After all, times and technology have changed. James B. Stockdale, at one point in his life, might have agreed with that criticism. Why, for example, does an aviator need a liberal education? Stockdale learned—in a most horrible fashion—differently. In September 1965 the aircraft he was piloting was shot down over North Vietnam. He was destined to spend 2,714 days imprisoned until his release in February 1973. He was to endure, heroically, three years of solitary confinement and more than a year of total isolation. A winner of the Congressional Medal of Honor, Admiral Stockdale has written eloquently that the philosophy course he took at Stanford as a graduate student did him more good during his captivity than any of the Naval Academy's technical subjects.[35] With Newman, Stockdale learned how important is a comrade you can depend on and how vital is the possession of "a mind which lives in itself"; with Cicero, that education is "a consoling refuge in adversity"; and with Hans Morgenthau (quote following), that today's mores may have nothing to do with the enduring things (as Koppel and C.S. Lewis have tried to tell us):

> Is not morality, so you might ask, a relative thing, the ever-changing result of environment and circumstances? If this were so, let me ask you, how do you explain that we cannot only understand the moral relevance of the Ten Commandments, originating in a different social environment and circumstances quite different from ours, but also make them the foundation for our moral life? How do you explain that the moral rules of Plato and Pascal, of Buddha and Thomas Aquinas are similarly acceptable to our intellectual understanding and moral sense? If the disparate historic systems of morality were not erected upon a common foundation of moral understanding and valuation, impervious to the changing conditions of time and place, we could neither understand any other moral system but our own, nor could any other moral system but our own have any moral relevance for us. It is only because we as moral beings have something in common with all other men—past and present—that we are able to understand, and make our own, the core of the moral systems of others.[36]

Liberal education, then, teaches us the long view. It can provide us with a moral and mental compass that can keep us ethically true despite the swirling tastes around us. How very much we need an ethical compass! How very much our professional officers need that compass! An Army general, Omar Bradley, summarizes the argument eloquently:

> We have grasped the mystery of the atom, and forgotten the Sermon on the Mount; we possess brilliance without wisdom, power without conscience. Our knowledge of science has outstripped our power to control that knowledge. We are a world of nuclear giants and ethical infants. We know

more about war than about peace; more about killing than about living; this is the twentieth century's claim to distinction and fame.[37]

If the military colleges are to be effective in their *de facto* role as the leader of ROTC; if the Norwich University of Alden Partridge is to continue to fulfill the role its founder dreamed for it; if ROTC is to be a meaningful and vital educational experience (as well as a useful training vehicle for cadets who would be lieutenants)—the central idea of genuine education must be inculcated in those who would bear the responsibility of a commission. They must know that the heart of leadership, in a word, is *integrity*. For without integrity—the fruit of an informed conscience (itself, in turn, the product of long thought and careful study)—the *soi-disant* leader is a hollow man. And without real leaders, there is no point to professional military education (the topic of the next chapter) and there is no possibility for an American military ethic—or for true faith and allegiance to this Republic.

NOTES

1. Quoted in Russell Kirk, *Decadence and Renewal in the Higher Learning* (South Bend, IN: Gateway, 1978), p. xi.

2. John Sloane Dickey, "Conscience and the Undergraduate," *The Atlantic*, April 1955: 32.

3. Lee S. Harford, Jr., "A History of Army Officer Training at Civilian Military Colleges 1819 to 1990"; paper prepared for the Norwich Symposium on the Future of Military Colleges held on 7 December 1990.

4. Quoted in Eliot A. Cohen, *Citizens and Soldiers: The Dilemmas of Military Service* (Ithaca, NY: Cornell University Press, 1985), p. 124.

5. John D. Kraus, "The Civilian Military Colleges in the Twentieth Century: Factors Influencing Their Survival." Unpublished Ph.D. Thesis; The University of Iowa, 1978, p. 386. His emphasis.

6. Quoted in Arthur A. Ekirch, Jr., *The Civilian and the Military* (New York: Oxford University Press, 1956), pp. 121–122. This is now reprinted by Ralph Myles, Publisher, in Colorado Springs (1972).

7. Two extreme examples of a *negative* view of West Point are K. Bruce Galloway and Robert B. Johnson, Jr., *West Point: America's Power Fraternity* (New York: Simon & Schuster, 1973) and Richard U'Ren, *Ivory Fortress: A Psychiatrist Looks at West Point* (New York: Bobbs-Merrill Co., 1974). Both of these volumes drip with venom and vitriol. The fact remains, however, that the U.S. Military Academy continues to attract some extraordinarily talented young people; it clearly qualifies as one of our country's finest educational establishments, some vindictive treatments to the contrary notwithstanding. See, for example, "West Point Makes a Comeback," *Time*, 4 November 1985: 28 and following.

8. Quoted in Forrest C. Pogue, *George C. Marshall: Education of a General* (New York: The Viking Press, 1963), p. 39. When other colleges abandoned their convictions and traditions, VMI stood firm. Although I am indebted to my own institution, Norwich University, for providing a secure haven in which to teach and write, I view with distress its decisions to incorporate a cluster of programs thatI see as antithetical to its historic purposes—all in the name of diversity. Institutions that try to be all things to all people may end up being nothing to anyone. VMI has long known what it is about. For discussion, see *Time*, 1 July 1991: 66, and Jere Real, "The Last of the Old Corps?" *National Review*, 6 August 1990: 22–23.

9. As long ago as 1959, Gene Lyons and John Masland were calling for increased substance in ROTC programs: "The better the student is, the more likely he is to be highly motivated by a challenging program expertly presented." *Education and Military Leadership: A Study of ROTC* (Princeton, NJ: Princeton University Press, 1959), p. 219.

10. Quoted in "Position Paper on the Status of the Senior ROTC Program in the USA," Reserve Officers Association of the United States [December 1990].

11. *Great Dialogues of Plato*, trans. by W. H. D. Rouse, and ed. by E. H. Warmington and P. G. Rouse (New York: New American Library, 1956), p. 443.

12. John Henry Cardinal Newman, *The Idea of a University* (New York: Longmans, Green and Co., 1947), p. 107.

13. McGeorge Bundy, "Were Those the Days?" *Daedalus* 99 (Summer 1970): 555.

14. Newman (above, n. 12), p. 107.

15. Kirk (above, n. 1), p. 296.

16. Ibid., pp. 294–295.

17. Ibid., p. 298.

18. John Leo, "The Academy's New Ayatollahs," *U.S. News and World Report*, 10 December 1990: 22. See also Scott Heller, "Colleges Becoming Havens of 'Political Correctness,' Some Scholars Say," *The Chronicle of Higher Education*, 21 November 1990: 1 and following; and Carolyn J. Mooney, "Academic Group Fighting the 'Politically Correct Left' Gains Momentum," *The Chronicle of Higher Education*, 12 December 1990: 1 and following. One of the clearest thinkers about academic issues—for example, he thinks that a professor can grade an examination without being a fascist—is Steven M. Cahn. See his two books: *The Eclipse of Excellence* (Washington, DC: Public Affairs Press, 1973); and *Saints and Scamps: Ethics in Academia* (Totowa, NJ: Rowman & Littlefield, 1986.) He is the editor of *Morality, Responsibility, and the University* (Philadelphia, PA: Temple University Press, 1990). Is there any institution that did not have debates about Allan Bloom, *The Closing of the American Mind* (New York: Simon & Schuster, 1987)? Bloom's book was not "politically correct"—much to its credit! Another book of interest is E.D. Hirsch, *Cultural Literacy: What Every American Needs to Know* (Boston, MA: Houghton Mifflin, 1987).

19. William Eckhardt, "The Military-Industrial Personality," *Journal of Contemporary Revolutions* 3 (Fall 1971): 74–75. To pursue this line of thought, see Eckhardt, et. al, "Militarism in Our Culture Today," *Journal of Human Relations* 15, No. 4: 532–537. And Eckhardt, "Psychology of War and Peace," *Journal of Human Relations* 16, No. 2: 239–249.

20. Walter Lippmann, *The Public Philosophy* (New York: The New American Library), p. 65.

21. Alfred Thayer Mahan, "Naval Education," *U.S. Naval Institute Proceedings* 5 (1897): 375.

22. Lippmann (above, n. 20) prefers this term; see Romans 2:15 to find reference to the idea that "what the Law requires is written in [our] hearts."

23. Army ROTC "ranks" college majors, with engineering subjects, science, and mathematics held in highest regard. When one of my sons, a four-year Army ROTC scholarship recipient, tried, in his junior year of college, to change his major from mathematics to philosophy, his petition, although approved with the highest endorsement of his professor of military science, was denied. As a mathematics major, he was not allowed to switch to a "lower" major. One of the reasons he was given for the denial was his excellent grade point average in mathematics; had he been failing, he was told, he would have had a better chance of having his petition approved! [Note: Some weeks after my son received the notice of denial of a change of major, he received a call from his professor of military science informing him that the Army would, after all, approve a change in major for him.]

24. Hans J. Morgenthau, "Epistle to the Columbians," *The New Republic*, 21 December 1959: 10.

25. Newman (above, n. 12), p. 40.

26. Quoted in Robert H. Bork, *The Tempting of America* (New York: The Free Press, 1990), p. 164.

27. Kirk (above, n. 1), p. 191.

28. Dickey (above, n. 2), p. 31.

29. Kirk (above, n. 1), p. 73.

30. H. G. Rickover, "Thoughts on Man's Purpose in Life and Other Matters," *U.S. Naval Institute Proceedings* 100 (December 1974): 72.

31. This is the essential meaning of Mt 16:25, Mk 8:35, and Lk 17:33. Although he doesn't accept military service as a form of "repayment" to our country, William F. Buckley, Jr. has raised some interesting points in his book *Gratitude* (New York: Random House, 1990). The military colleges and ROTC are founded on the idea that we, in fact, owe our country something—an idea that Socrates would have endorsed. See Plato's dialogue *Crito*.

32. Newman (above, n. 12), pp. 157–158.

33. Cicero is quoted in John Bartlett, *Familiar Quotations*, ed. by Emily Morison Beck (Boston, MA: Little, Brown, 1968), p. 110.

34. John Milton, "Of Education," in his 1644 letter to Mr. Hartlib in *Areopagitica and Other Prose Works* (New York: E. P. Dutton & Co., 1946), p. 46.

35. James Bond Stockdale and Sybil Stockdale, *In Love and War* (New York: Harper & Row, 1984); and J. B. Stockdale, "The World of Epictetus," *The Atlantic*, April 1978: 98–106 (a very worthwhile piece for those who prefer mathematics to philosophy for ROTC cadets!); George C. Wilson, "A Former POW's Search for Ways to Teach Courage," *Washington Post*, 9 April 1978: C1–C2; "This Prof Learned the Hard Way," *Time*, 19 February 1979: 98.

36. Morgenthau (above, n. 24), p. 10.

37. Quoted in Sister M. Claire Madeleine, "The Teacher As Thinker Before His Class," *Liberal Education* 48 (March 1962): 33. In "The Pen and the Sword: A

Heuristic Study of the Compatibility of Liberal Education and Military Training on the College Campus" (unpublished monograph, 1980), available in the libraries of Norwich University and Virginia Military Institute, I sought to deal with a number of the problems I have again explored here. In *Military Review* 70 (November 1990), Stephen Danckert makes a number of good suggestions about improving ROTC (pp. 75–77), including having the professor of military science at each ROTC unit educated to the doctoral level and having him or her designated a permanent professor. There are numerous problems with that idea (Ph.D.-level education may be generally impracticable on such a scale, and colleges ought to control their own promotion and tenure schedule), but no one can quarrel with the idea of raising ROTC academic standards. Such proposals as mine in "The Pen and the Sword" and those of Captain Danckert at least put the emphasis in the right places. See also "Competence and Character," *U.S. News and World Report*, 27 May 1991: 36, about General Schwarzkopf's comments at West Point.

7 PME: PROFESSIONAL MILITARY EDUCATION

Wisdom is better than weapons of war, and one mistake can undo many things done well.

—Ecclesiastes 9:18 (NEB)

Nothing so comforts the military mind as the maxim of a great but dead general.

—Barbara Tuchman, *The Guns of August*, 1962

The scene is the main ballroom of the Officers' Club at Fort Leavenworth, Kansas. It is 20 January 1988, and Congressman Ike Skelton of Missouri, Chairman of the Panel on Military Education of the Committee on Armed Services of the House of Representatives, prepares to receive the testimony of Major General Gordon R. Sullivan, Norwich University alumnus and Deputy Commandant of the Command and General Staff College (and, in 1991, Army Chief of Staff). At the end of his testimony, General Sullivan offers this opinion: "I would just say in summary that we are ever mindful of the lessons of history. We understand our role in the system: that is, in our Government, and we . . . appreciate the emphasis that you are placing on officer education because we are ever mindful of the statement of a British officer, when he said, 'The nation that insists on drawing a line between fighters and thinkers is doomed to failure.'"

Congressman Skelton replied: "To complete the quote, 'You will have your fighting done by fools and your thinking done by cowards.'"[1] Almost every lieutenant will, in time, become a lieutenant colonel, or colonel, or general: or he will earlier leave the service. His education, therefore, must be a matter of fundamental concern to the American public. Brilliant lieutenants may not become brilliant colonels; and brilliant colonels may not become brilliant generals, a fact recognized centuries ago by the Comte De Saxe: "I have seen very good colonels become very bad generals. I have known others who were great takers of villages . . . but who, outside of that, were not even able to lead a thousand men in war, who lost their heads completely and were unable to make any decision" (*Reveries*, 1757). But dumb generals are always dumb generals. Historians debate the extent to which dumbness struck General Douglas Haig (1861–1928) of World War I. But at least one of his detractors, David Lloyd George, is supposed to have said of the British general that he was "brilliant to the top of his army boots."[2]

Through incompetent action or dissolute character, dumb generals kill the very troops of whom they have charge. It is therefore clearly in the national interest that investigation into the nature and content of professional military education (PME) be conducted by appropriate agencies. We know relatively much about ROTC and West Point; we know surprisingly little about the education of those people likely to rise to the highest ranks in our armed forces. This chapter inquires into the purposes and programs of PME, discussing the men and institutions associated with the arcane and abstruse field of professional military education.

Before examining PME schools and the general philosophy of PME, one should know how PME is defined by the Joint Chiefs of Staff. The official definition of PME (taken from a Joint Chiefs of Staff paper [*Military Education Policy Document*, CM 344-90, 1 May 1990]) is this:

> Professional Military Education (PME) provides individuals with the skills, knowledge, and understanding that enable them to make sound decisions in progressively more demanding command and staff positions within the national security environment. PME has as its premier theme the employment of combat forces, with strategy being increasingly emphasized at the intermediate, senior, and general/flag officer levels. It addresses the military, political, economic, social, and psychological dimensions of na-

tional security with varying degrees of emphasis on the planning and conduct of war, Service organizations, joint and combined operations, force employment and deployment concepts, and military leadership.

If not the most succinct definition, that explanation does at least convey most of the objectives of American professional military education.

I

The United States Air Force Academy, the United States Coast Guard Academy, The Citadel—these undergraduate military colleges are fairly well known to the reading public and to those with interest in national security affairs. Even the initials of some of our undergraduate military schools—VMI, the USMA (West Point), the USNA (Annapolis)—are similarly well known. But what of the USAC & GSC, or of the ICAF, or of the AWC? The Army's Command and General Staff College, the Industrial College of the Armed Forces, and the Air War College are vital parts of American PME. Yet they are little known. Since the time at least of Elihu Root (1845–1937), who reorganized the command structure of the Army under President McKinley,[3] some military analysts have concentrated their efforts on establishing a PME system in some ways reminiscent of the Prussian system, yet one wholly in keeping with the democratic traditions and character of our Republic. In the lives of our professional leaders in politics, in diplomacy, in law, in medicine, which is of the greater importance: their undergraduate experience or their graduate or professional experience? To be sure, both are important, and one cannot have the latter without the former. Still, the professional (post-collegiate) education of our armed forces senior officers has been, in the main, neglected for decades, and some might argue for centuries.

Few activities in life are conducted independent of political exigencies. Even PME must be developed and delivered within a political context. Yet this is not intended to malign the Skelton Committee. Ike Skelton was hardly canvassing votes in his role as chairman of his committee. Although some military officers might quarrel about the exact nature of some of the recommendations made by the Skelton Committee, few officers will withhold praise of Skelton and his committee, for it is widely and correctly recognized that Congressman Skelton and his committee, through hard work and thoughtful analysis, were addressing a matter of perennial importance to our country and its citizenry. For as we do not want dumb lieutenants, we most assuredly do not want dumb generals. If ROTC, West Point, and OCS are important to us, how much more so are the military schools in which carefully selected senior officers are educated? The Skelton Committee of 1988–1989 performed substantial, if little-known, service to the country in their efforts to review and to reform PME.

But even they had to operate within a given political structure. Today, that structure is known by a single word: "jointness." Congress in the fall of 1986 passed the Goldwater-Nichols Department of Defense Reorganization Act. That Act significantly strengthened the powers of the Chairman of the Joint Chiefs of Staff (JCS). Additionally, the act improved the powers of the commanders of the unified and specified commands, although these subjects are beyond the confines of this book. More to the present point, Goldwater-Nichols addressed a subject of enduring concern to many critics of our armed forces: the alleged excessive loyalty to one service—rather than to the nation as a whole—of some officers. To be sure, officers are promoted in and rewarded by their own service. In an effort to remedy service provincialism, Goldwater-Nichols required that fifty percent of joint-duty positions above the rank of captain (Navy lieutenant [0-3]) be filled by officers selected for the joint specialty and, with some exceptions, that no officer could be promoted to general (or flag) rank unless he had already served in a joint-duty assignment.[4]

Response to this element of the Goldwater-Nichols Act was mixed. Those who formerly worked in such joint assignments (sometimes referred to as "purple-suiters"—as opposed to green-suiters [Army] or "blue-suiters" [Air Force]) supposedly lost out in competition for promotion in their own services because of their purple suit service. Goldwater-Nichols, therefore, specifically addressed that concern: Purple-suit service is said to be highly regarded and warmly smiled upon by the gods of promotion. Still, there are senior officers who question the wisdom of what they see as now excessive commitment to pooled service operations. Each of our services is different, after all, they might say, and it is possible to overemphasize similarities and centralization.

Be that as it may, Goldwater-Nichols has had a decidedly healthful effect on what are known as Joint Service Schools, some of which existed a long time before the 1986 Department of Defense Reorganization Act. It may be worth listing these schools, from which list I will omit at present the chief institutes with which we will be concerned later in the chapter:

- Uniformed Services University of the Health Sciences (Bethesda, MD)
- Defense Institute of Security Assistance Management (Wright-Patterson AFB, OH)
- Defense Intelligence College (Washington, DC)
- Defense Mapping School (Fort Belvoir, VA)
- Defense Equal Opportunity Management Institute (Patrick AFB, FL)
- Defense Systems Management College (Fort Belvoir, VA)
- Defense Information School (Fort Benjamin Harrison, IN)
- Defense Foreign Language Institute (Presidio of Monterey, CA)
- Defense Language Institute, English Language Course (Lackland AFB, TX)
- Defense Resources Management Education Center (Monterey, CA)
- Joint Military Packaging Training Center (Aberdeen Proving Ground, MD)

Of course this list of Joint Service Schools does not incorporate the joint training done by certain military skills centers. For example, the U.S. Army Airborne School at Fort Benning trains members of all the armed services in parachute techniques, not just soldiers; the Ranger School, similarly, trains principally soldiers but is open to qualified members of other services as well. Examples could be multiplied, but each is concerned with military skills. Here we are for the most part concerned with "higher education."

At Fort Lesley J. McNair in Washington, DC, is the National War College, which opened in 1946 after a number of American leaders became convinced that an interservice school was required at the zenith of the military educational system. The Navy has had a war college in Newport, Rhode Island, since 1884, and the Army's war college has existed since 1901 in Carlisle Barracks, Pennsylvania. The new Air Force created its own war-college at Maxwell AFB in Montgomery, Alabama, in 1946. Also at Fort McNair is the Industrial College of the Armed Forces, which is essentially concerned with logistics and supply. It had been set up in 1924 as the Army Industrial College after questions arose about the availability of supplies for U.S. troops during World War I. It ceased operations during World

War II, but it was reconstituted in 1946.[5] These, then, are the five senior-level war colleges of the PME system in the United States:

- National War College (Fort McNair, Washington, DC)
- Industrial College of the Armed Forces (Fort McNair, Washington, DC)
- Army War College (Carlisle Barracks, PA)
- College of Naval Warfare (Newport, RI)
- Air War College (Maxwell AFB, Montgomery, AL)

Students of the PME system are understandably confused, at first, by the complexity of the system. For example, the National War College is one of three schools run by the National Defense University. (The other two under its aegis are ICAF [Industrial College of the Armed Forces, previously mentioned] and the Armed Forces Staff College [alluded to in the following]). Similarly, at Maxwell Air Force Base, one finds not only the Air War College (see preceding list) but also the Air Command and Staff College (see following list), as well as a number of other educational enterprises collectively known as the Air University.

In addition to the five senior-level war colleges, there are five intermediate-level colleges:

- the Armed Forces Staff College (Norfolk, VA—under National Defense University)
- the Army Command and General Staff College (Fort Leavenworth, KS)
- the College of Naval Command and Staff (Newport, RI)
- the Air Command and Staff College (Montgomery, AL—under Air University)
- the Marine Corps Command and Staff College (Quantico, VA)

Of these ten PME schools, the National War College, ICAF, and the Armed Forces Staff College are considered "joint," or multiservice schools. All the colleges, however, admit officers from sister services, selected government civilians, and certain foreign officers. The joint schools, however, are not directly run by one uniformed service.

There is one other school or program conducted at Fort McNair that is of interest here. That is the Capstone course for newly minted general/flag rank officers. It is sometimes referred to as "charm school" by lower-ranking officers, most of whom would be delighted to accept the instruction (and the stars) associated with that program!

Enrollments in these colleges will be of interest. Capstone meets four times a year; enrollment is about thirty-five officers per class. The Armed Forces Staff College meets twice a year; its enrollment is about 245 officers per class. The other colleges conduct programs lasting for about ten months—an academic year. The National War College enrolls 118; ICAF enrolls 175; the Army War College, 215; Navy War College, 173; Air War College, 183; the Army Command and Staff College, 765; the Navy Command and Staff College, 162; the Air Command and Staff College, 450; and the Marine Corps Command and Staff College, 146. (These figures are for annual U.S. officer enrollment.)[6] Approximately 2,000 American officers attend these ten (eleven, with Capstone) PME schools annually. They are generally regarded as the rising stars of the U.S. military. Although the PME schools have made substantial progress, certain problems persist.

Two other military colleges should be mentioned here. At Monterey, California, is the U.S. Naval Postgraduate School (NPS). Established in 1909 to meet the advanced educational needs of Navy officers in marine engineering, the NPS has evolved into an academic institution of note, particularly in technical fields. Designed primarily for lieutenants (0-3 rank) and lieutenant commanders (0-4), primary emphasis at the degree-granting NPS is placed on scientific and technical studies, although courses and programs in other fields (e.g., national security) are offered. (The war colleges, by contrast, are primarily concerned with national military strategy and with national security strategy—about which more will follow.) At Dayton, Ohio, in 1919, the U.S. Air Force Institute of Technology (AFIT) officially opened. This degree-granting institution is, as its name clearly suggests, primarily concerned with engineering and scientific studies. Technically a part of Air University in Montgomery, most of AFIT's activities are conducted by its schools of Engineering, of Systems and Logistics, of Civil Engineering and Services, and by various research institutes in such fields as operations and management. While NPS and AFIT are certainly valuable components of the American professional military educational system (and, in fact, offer instruction in myriad fields), they are chiefly aimed at and interested in scientific, managerial, and operational instruction.

II

The first difficulty that any system of PME must surmount is a virulent anti-intellectualism, not only in the U.S. military, but in American society.[7] Military officers are, of course, drawn from American society. If the parent society sees little point in developing the qualities of erudition, that is likely to be true as well with its officer corps. This is not to impugn the intellect of the officer corps as a whole; it is, however, emphatically to question the value customarily assigned by military leaders to exactly the kind of education we discussed in the last chapter. Is there a point, in the imaginations of our highest military leaders (or, indeed, in the imagination of the general American public), in developing the tastes and attitudes normally associated with the liberal arts? Is there purpose in reading, in thinking, in conversing—beyond the level that ostensibly separates us from the beasts? Not everyone would readily agree!

Although the intermediate-level PME schools are principally concerned with the tactical and military arts, the senior service schools (such as the Air War College) offer what in some cases can be correctly described as graduate-level education in such fields as The Senior Leader, National Security Decision Making, Military Strategy Analysis, Forces and Capabilities, Regional Security Analysis, and International Security Studies. Moreover, during the ten-month-long course, students at the war colleges choose from a large menu of advanced and elective courses including Napoleonic Warfare, Space Issues, The Economics of Defense, Executive Editing, Media Relations for Senior Officers, The American Civil War, and Command and Conscience (which I created for the Air War College). The announced mission of the U.S. Air War College is "to improve the Air Force contribution to national security through joint education and senior-leader development." Of course airpower is the main concern at Maxwell Air Force Base, whereas ground power is the principal focus at Carlisle Barracks, home of the Army War College. But war colleges are concerned with the education, the development, and the intellectual enrichment of those who will soon be leading the services. In a nutshell, the idea of the war college is to produce fighters who can write, think, and speak effectively. Those chosen to attend a war college (usually considered to be a career-enhancing opportunity) have been very successful in their military field. Now they are expected to grow beyond "heavy equipment operation" (as some Air force pilots self-deprecatingly describe their work) and think about international strategy. Whereas before war college students *trained*, now they are expected to be *educated*. Attending lectures and seminars, perusing fairly substantial reading lists, writing papers and book reviews, and studying for exams are new to colonels twenty years or more out of college. But the transition from tactics to strategy, from reading training manuals to writing analytical papers, is critical for their development and their services' future. The war colleges exist to help produce not just warriors, but thinkers—true soldier-scholars. Some senior officers have trouble making this critical career transition. Not long ago in the barbershop at Maxwell Air Force Base, I heard two Army lieutenant colonels discussing upcoming papers required of them. "Hell," said one, "I thought I was supposed to be a fighter—not a writer."

That opinion, exactly, was shared by Rear Admiral F. H. Ramsey who, in endorsing the unfavorable fitness report of an officer in 1893, wrote that "It is not the business of a naval officer to write books." The officer receiving that unfavorable fitness report was Alfred Thayer Mahan (1840–1914).[8] Even so ardent an admirer of the western military as Basil H. Liddell Hart once said that "the Army, for all its good points, is a cramping place for a thinking man."[9] I could have made a similar judgment after only one conversation with my old company commander from West Point, Goppy Dolan. Liddell Hart was right: those involved in PME are, as a matter of simple fact, going against the grain. Our society, generally speaking, is anti-intellectual. That is not to say that it is stupid; Americans (and their officers) command a great deal of factual knowledge. To say that we and our military are "anti -intellectual" is to suggest only the obvious: that the kind of liberal education described in chapter 6 of this volume is not customarily held in high regard by our technological, empirical, quantitative, and pragmatic culture. There is hardly anything wrong with being scientifically and quantitatively literate; indeed, it is a desideratum. But numbers alone, important as they are, are not the sole content (or the soul content!) of statecraft and strategy, of politics and geopolitics, of ethics and economics. And it is these types of subjects that our senior-level PME schools seek to teach our rising leadership. Effective PME, therefore, adds to and burnishes a substantial undergraduate education; it also depends upon senior leaders having had such an effective education earlier in their careers. Effective PME demands effective junior-level education.

If our senior leaders are not senior readers, they will have great difficulty in developing the kind of clear, coherent national security strategy our country needs. The absurd comment that intellectuals are always ivory-tower scholars without so much as a shred of common sense is a useless *ad hominem* argument. Of course, intellectual soldiers must have knowledge of the real world. (One thinks of Augustine, Machiavelli, Hobbes, and Luther—how much more "real world" can one be?) The Skelton Report suggests that George Marshall (a VMI alumnus) was a great strategist—"the result of Marshall's (1) being taught to think broadly and (2) taking the time to think extensively and [to] reflect on that reading."[10] What I mean, then, by intellectual, is one who cherishes the value of learning, particularly of academic (as opposed to popular) reading, writing, and discourse; one who values disciplined reasoning and seeks to apply the fruits of his academic learning to a variety of worthy endeavors.[11] That definition, simple as it is, does not absorb those people or officers who rarely, if ever, read any serious books in history, politics, philosophy, theology, economics, and great literature; who do not keep a personal journal or seek occasionally to express themselves in writing in a serious forum—there is no better way to bring coherence to your own thoughts than by organizing it in writing; or who flee in mental panic from the first sign of a serious conversation (as opposed to social small talk or debate about which team is likely this year to win the World Series). That is all I mean by intellectual. And yet it is too much for too many.

If we think Liddell Hart seems cruel in saying that "the only thing harder than getting a new idea into the military mind is to get an old one out,"[12] consider this animadversion penned after World War I by H.G. Wells in his popular book, *The Outline of History*:

> [T]he professional military mind is by necessity an inferior and unimaginative mind no man of high intellectual quality would willingly imprison his gifts in such a calling; nearly all supremely great soldiers have been either inexperienced, fresh-minded young men like Alexander, Napoleon, and Hoche, politicians turned soldiers like Julius Caesar, nomads like the Hun and Mongol captains, or amateurs like Cromwell and Washington: whereas this war [WWI] after fifty years of militarism was a hopelessly professional war.[13]

Some years ago, Maureen Mylander studied general officers. In her study, she of course investigated "schools for generals" (the war colleges), although I am not sure that she appreciated the fact that one did not then and need not now have to be a war college alumnus to be a general. Nevertheless, in view of Wells's stinging rebuke three-quarters of a century ago about the military mind, one would think that, by the end of America's participation in the Vietnam War more than fifty years later, there would have been a great change in PME. Here is Mylander's assessment:

> As they prepare generals-elect for future political and military roles the war colleges seem to have all the advantages: famous graduates, prestige, impressive physical facilities, cohesive student bodies, illustrious lecturers, excellent libraries, field trips and cooperative degrees. The schools have upgraded faculty, made innumerable administrative changes, and tinkered ceaselessly with their curricula. Yet in trying to generate an atmosphere conducive to creative, original thinking, the war colleges have failed.[14]

More than fifteen years after Mylander's assessment of the war colleges came Michael Massing's brief analysis of the Army War College, which, he said, "suffers from institutional inertia." Massing also criticized the Army War College in this regard: "Until senior officers understand the social roots of military conflict, they cannot hope to resolve it. This, in fact, is the primary lesson of Clausewitz, who wrote that war is an extension of political conflict. Apparently, this notion hasn't yet pervaded the war college's consciousness where the Third World is concerned."[15]

To the extent that American society resists the attributes of the intellectual I described, the war colleges must seek to overcome, to alter, to amend that tendency. But consider that the staff, faculty, and students of the war colleges, or most of them at any rate, have been socialized by exactly that process that must now be changed—by them!—if we are to produce intellectual colonels and generals. Too many officers in today's military are entrepreneurs[16] or military courtiers.[17] If our leaders in the future cannot unlearn the bad ideas of the past, while preserving the treasures of our heritage (*and* being able to differentiate between the two!), one fears that our men and our institutions will suffer from the kind of rigor mortis that the Comte De Saxe so brilliantly described in his *Reveries* of 1757:

> Many commanding generals only spend their time on the day of battle in making their troops march in a straight line, in seeing that they keep their proper distances, in answering questions which their aides de camp come to

ask, in sending them hither and thither, and in running about incessantly themselves. In short, they try to do everything and, as a result, do nothing. They appear to me like men with their heads turned, who no longer see anything and who are only able to do what they have done all their lives, which is to conduct troops methodically under the orders of a commander. How does this happen? It is because very few men occupy themselves with the higher problems of war. They pass their lives drilling troops and believe that this is the only branch of the military art. When they arrive at the command of armies they are totally ignorant, and, in default of knowing what should be done, they do what they know.

The chief benefit of liberal education is that it endows its children with the perspective of the ages. A liberally educated man or woman is not, then, 25 or 35 or 45 or 55; he or she, in reality, is hundreds or even thousands of years old. Liberal education grants *perspective* and *discernment*. Thus, the liberally educated know, generally, what should be done; thus they are empowered, by knowing what ought to be done, to be free to determine what can be done, and thus to do what must be done. One cannot lead until one knows where to lead. The magnificent allegory of the cave (in Book VII of the *Republic* by Plato) is thus every bit as military as any Army field manual.

III

There is a story, perhaps apocryphal, that as the Swedish diplomat Axel Oxenstierna (1583–1654) lay dying, he was asked by his son to provide a distillation of his diplomatic wisdom, lest it be lost with his perishing. The elder diplomat responded: *Quantula sapientia regitur mundus.* ("With how little wisdom is the world governed.") To put it crassly, what he was saying amounts to this: people in authority are stupid; they govern stupidly; so don't expect too much. The French statesman Talleyrand (1754–1838) provided a similar apothegm: "Above all else," he counseled, "not too much zeal!"

Can it be that a principal reason for military failure is excessive zealousness coupled with stupidity? Is our commanders' collective lust for glory complemented by diminished intelligence? Can it be true that the higher we go in the military hierarchy the less wisdom we find? "What is it about the American officer corps that seems to produce officers who fail to succeed [*sic*] on the battlefield?"[18] that tendentious critic of the American military, Richard Gabriel, has asked. He is persuaded that "[M]any American officers are amateurs" and that "The poor quality of leadership at the lower levels is recognized by soldiers in the ranks." (Of course, this is the same Gabriel who was predicting that, in the Gulf War of 1991, the United States would have "8,000 to 10,000 killed, 30,000 to 50,000 wounded, and 40,000 [psychiatric cases]."[19]

But perhaps Gabriel has missed the point. Is it *American* officers who are, or so it is argued, incompetent (which, by the way, he defines simply as "the inability of military leaders and forces to avoid mistakes which, in the normal course of things, should and could be avoided"[20]), or is there a tendency of military leaders (not country-specific) to be incompetent? Strauss and Ober recently suggested of ancient military leaders that they "often erred because of a failure (which need not mean a lack) of intellect, rather than a failure of talent, energy, or ambition .... They lost because they avoided the painful but necessary process of thinking through and trying to overcome unexamined assumptions, opinions, and prejudices about themselves and about their enemies."[21] One can argue, then, in the light of that analysis—which differs from Gabriel's in that it examines ancient leaders of ancient states—that the more truly educated one is, the less likely he is to be myopic and provincial, hence incompetent. If one of the traditional fruits of education is the provision to the student of broad, Catholic horizons (and, presumably, increased analytical powers), we ought to be able to avert incompetence, or much of it, through the power of an educated officer corps. As Strauss and Ober write, "[V]ictory is not easy, but it is possible if one is willing to think through the most cherished assumptions."[22]

If a major impediment to developing leaders of competence and character is the inadequacy of their education, we have a hook upon which to hang the reformer's hat. If, however, the problem is systemic—that is, if the entire military system is somehow diseased and debauched—efforts at improving PME are as misdirected and counterproductive as the entire military itself. If the *Titanic* is about to sink, one does not worry about revising the curriculum of its officers.

Some years ago, Norman Dixon set out "to explain how a minority of individuals come to inflict upon their fellow men depths of misery and pain virtually unknown in other walks of life."[23] His conclusions, if correct, underscore the essential futility of trying to improve PME. Wrote Dixon:

> Military organizations make for military incompetence in two ways—directly, by forcing their members to act in a fashion that is not always

conducive to military success, and indirectly, by attracting, selecting and promoting a minority of people with particular defects of intellect and personality.[24]

Thus, there is a kind of chicken-and-egg problem to be resolved. Is it an incompetent military that attracts incompetent aspiring leaders, or is it those incompetent aspiring leaders who create an incompetent military? If that seems too severe, consider Dixon again:

> [M]ilitary organizations may have a particular propensity for attracting a minority of individuals who might prove a menace at high levels of command, and . . . the nature of militarism serves to accentuate those very traits which may ultimately prove disastrous.[25]

Of course, there is only a "minority" of people at high levels of command in the first place. So there is no doubt that, according to Dixon, the military services have a profound psychological problem. Military incompetence, in his judgment, derives not from lack of education proper and not from lack of intelligence as such, but rather from " susceptibility to cognitive dissonance [inability to consider alternatives], the tendency to pontificate and the inability to adjust the riskiness of decisions to the real situation." These properties, in turn, he suggests, "are a product of such neurotic disabilities as extreme anxiety under stress, low self-esteem, nervousness, the need for approval and general defensiveness."[26] A general's personality, suggests Dixon in a chapter entitled "Socialization and the Anal Character," bears the hallmark of his early toilet training.[27]

Dixon discusses the military's tendency to be authoritarian and suggests that military officers are inclined in that direction. No one seriously challenges that, and serious students of the American military from Samuel Huntington on would say as much—and none of it is pejorative. But when Dixon contends that "the events of Third Ypres [World War I] . . . constitute the acting out of an anal fantasy of impressive proportions,"[28] it's time to call a halt. Perhaps the best rejoinder to Dixon is the one recently provided by Cohen and Gooch:

> It would be easy to dismiss Dixon's theorizing as simply the kind of thing that gets psychoanalysis a bad name. In fact, a little reflection on the reality of military history provides ample refutation of his theories. Though many of the personality traits he identifies do seem to be present in some of the more notable cases of failure, it is by no means obvious that, by the same token, they are not also among the mental baggage of history's successful

generals. Many great commanders have had their share of obsessional jealousy, mental rigidity, and authoritarianism . . . .

Even more perplexing . . . is the case of the commander who fails at one time but creates a remarkable victory at another. The Douglas MacArthur who so utterly misjudged the likelihood and imminence of China's entry into the Korean War was, after all, the same Douglas MacArthur who first conceived and then implemented the Inchon landing in the teeth of strong opposition from his subordinates. Was he struck by a sudden attack of anal-retentiveness sometime between June and October 1950? It seems unlikely.[29]

This is not the forum to attempt a dissertation on politics and psychology. The studies done, predictably, are of various quality. But what concerns us here is the charge that our senior leaders are so subject to their own dark passions that, frequently if not always, they will prove inefficient or unethical before or during or after the battle. If Abraham Lincoln, as one analyst wrote, "could have been helped to understand the nature of the anguish produced by his inner conflicts . . . perhaps there would not have been any need for the bloody killings of the Civil War." But, as Arthur Schlesinger, Jr. points out so well:

> The historian can only find breath-taking the certitude with which [certain psychiatrists] announce their historical verdicts. They seem to find no real problems in public affairs, only projected problems. But—no kidding, fellows—the Civil War was not fought because of Lincoln's neuroses . . . . No amount of psychiatric exorcism could have done away with the intractable problems existing quite outside Lincoln's psyche in 1861.[30]

Schlesinger quotes Jefferson at one point: "Sometimes it is said that man cannot be trusted with the government of himself. Can he, then, be trusted with the government of others? Or have we found angels in the form of kings to govern him?" Schlesinger observes that "Replacing *kings* by *psychiatrists* does not greatly improve the sentence."

Nor, one contends, will the American military ethic be preserved or enhanced by our replacing *senior military leaders* by *psychiatrists.* To be sure, there are subjective problems with military leaders. Some military leaders are mentally and morally unfit for command. So are some clergymen, some professors, some doctors, some lawyers. Yet we do not, normally, propose to burn down all churches because of the greed or depredations of some priests or ministers; we do not raze all universities because of the trumpeted nonsense of some academics; nor do we level hospitals or law schools because of the eccentricities of some medical or legal practitioners. Why then should we objectively blame the military profession for the subjective inadequacies of some of the practitioners of that art and science? In the end, Cohen and Gooch are right in suggesting that "[t]he causes of military misfortune are complex," although there are, they say, three kinds of failure: "failure to learn, failure to anticipate, and failure to adapt."[31]

IV

Although the system of PME in the United States can be criticized, one must simultaneously suggest that few institutions engage, as vigorously as do the war colleges, in self study, seriously attempting to improve their atmospheres and their offerings. This book itself would have been impossible without the opportunities for study, reflection, and writing afforded to me by the Air War College. The American military is undoubtedly the most studied and written-about armed force in the world. It is remarkably—spectacularly—open to self-criticism. Some of its most vitriolic critics, by the way, are repeatedly asked to lecture at the PME schools. If this is not exactly a spirit of "hit me again!" it does clearly show that the military and its highest schools are ready to listen, eager to learn, and anxious (perhaps, as we will discuss chapter 9, too anxious) to respond.

Having said that, one must nevertheless consider the criticisms leveled against the war colleges, one controversial source of which is a recent book by Martin van Creveld, entitled *The Training of Officers*. Van Creveld points out, with regard to war college education, that entrance into the five top military schools depends upon the candidates' general career patterns, not upon their performance, for instance, in a competitive examination. Some students entering the war colleges are underprepared in two ways: academically, for they lack the undergraduate knowledge base required to pursue graduate level studies; and attitudinally, for they lack the commitment to spend the time and energy necessary to obtain the type of education that the war college tries to provide them. As van Creveld suggests, the length of study at the War Colleges (ten months) is simply inadequate to provide both rigor and time for reflection. To expect a longer commitment from the students, however, simply doesn't square with the competitive realities of today's military; those in the war colleges, removed as they are from the mainstream command positions of their services, may actually fall behind their peers on the ladder of advancement.

Van Creveld criticizes the war colleges for having inadequate jointness. Although each of the three service war colleges (Air, Army, and Navy) has a collection of officers from all the services, those schools are, basically, the highest educational institution in their particular branch. The National War College and ICAF, however, are joint schools at the senior level of PME, and the Armed Forces Staff College is a joint school at the intermediate level. I strongly disagree with van Creveld's implication that all PME schools should be joint. The Air Force, Army, and Navy, in my view, must have their own senior-level war college. Increasingly, for example, the Air Force "thinks" at their war college; there must be a place of rigor where each of our services can think through its doctrine and help to effect a bonding among its chosen senior leaders of the present and future. This hardly rules out "seasoning" each war college class with officers from other services, with civilians, and with foreign officers. But true jointness should come at the real acme of the PME system.

Another criticism made by van Creveld is that PME schools lack rigor. That complaint, largely justified today, was echoed by Maureen Mylander some years ago when she complained bitterly about the military briefing that (although it has its place in the field) is absurdly out of place at the level of a war college. As Mylander wrote:

> The briefing is both symbol and symptom of what is wrong with the war colleges: preoccupation with form over content, and *a belief that wisdom equates to knowledge* [my emphasis]. A war college education, like the briefing, is a distributive rather than cerebral process. The schools permit their charges to range over vast expanses of subject matter but keep the herd moving so fast that there is scant time for intellectual grazing.[32]

Some war college students resent examinations. They do not want to be tested, and they think that exams are beneath their professional dignity. One might concede that certain types of exams are, in fact, inadvisable in a setting of that sort. These are professional people who have risen to the virtual top of their chosen field. At the same time, however, there is an old military maxim, the wisdom of which applies here: "The organization does well only those things that the boss checks." So it is with examinations, which offer a carrot to overachievers and a stick to the underachievers. Before the Skelton Committee made its recommendations, only the Navy War College (of the five senior-level PME schools) had graded examinations and a distinguished graduate program.[33] The notion that competition should be suspended during academic time spent at the PME schools is one I do not agree with. Some Air Force officers in seminars at the Air War College argue heatedly that strong scholastic attainment need not correlate with military success. But, in reply, one may suggest that academic success is an indicator (and an important one) of the kind of stuff we seek as we educate (not train) strategists. Indeed, the very nature of war college education is to help its students think more clearly, reason more critically, write more cogently, speak more convincingly. If those vital talents can be effectively developed and taught, they can also be fairly graded. And should be. Students at war colleges should be able to fail. The effective captain is not necessarily an inchoate good colonel. The good colonel, as de Saxe observed, is not necessarily an inchoate good general or admiral. The academically accomplished war college student is, of course, no sure-fire great general of the future. But add to the fact that he or she is unlikely to be at the war college without having performed very creditably in the field, the pleasant realization that this is a remarkably intelligent and analytical person and there is a good bet for someone who may, in fact, be a true "strategist" of the future.

Van Creveld strongly suggests that the senior PME schools be authorized, after attaining the appropriate standards, to grant the Ph.D. (or at least an M.A.) in military studies or military science. One can certainly argue about this, but I must question whether it is the purpose of PME to grant an academic credential. One, instead, prefers to think that a small body of officers might (as at present) be chosen to pursue advanced degrees at known civilian institutions. PME is not intended to be a graduate school, although it should have graduate-education rigor. The officers who are selected to take advanced degrees might be aimed, ultimately, at teaching at the federal service academies. Indeed, a service branch given to academic attainment, the professoriate, seems logical, for they exist in law, in medicine, and in religion. But the purpose of PME is not to produce teachers; it is to produce strategists.

Finally, van Creveld criticizes the faculty at the war colleges. Many of the military faculty at the war colleges are, in essence, draftees, selected to teach at the war college after their completion of the student curriculum there. For obvious reasons, that is unsatisfactory. We return once more to the idea of a corps of professional military educators, from whose ranks some of the faculty at the senior PME and intermediate-level PME might well be drawn. A greater problem exists with civilian faculty, who must be highly competent in their fields, organized, fair, and enthusiastic. In a word, they must fit into the PME schools. That is not an easy thing for some civilian educators, however sharp they might be. The war colleges are populated with confident, bright, energetic achievers. They can be very candid in their critiques, oral and written. If a given professor is overly sensitive, that person may have problems of adjustment in the rather freewheeling professional seminars at the war colleges. Not in the least is this intended to be a pejorative comment, either toward war college classmembers or civilian professors at the war colleges. It is merely to suggest that one must recruit war college professors carefully. There is no substitute for high professional competence, and no ideological test ought to be applied—save that of common sense: some elements of a war college education are legitimately classified. It is therefore reasonable to presume that the professor ought to be able to meet the criteria of a security clearance. Of course, one agrees strongly with van Creveld's call for an excellent civilian war college faculty. One can point to the Navy War College as setting the pace in that regard.[34]

There is no question that the Skelton Committee did yeoman service in reviewing the war colleges' programs and in making a series of wise recommendations for improvements. At the heart of their recommendations was this one: There should be a National Center for Strategic Studies, at which national security strategy[35] would be taught. At ICAF (which is principally concerned with mobilization and logistics) and at the three service senior-level war colleges, the academic concentration would be on national military strategy.[36] The intermediate level PME schools would concentrate on combined arms operations and joint operational art—the practical business of the military services. The National War College in Washington, DC, would be converted into the National Center for Strategic Studies under this scheme. The Report of the Skelton Committee, in my view, is a model of disinterested analysis. Its recommendations are prudent and practicable. A number have already been implemented by the war colleges. Much remains to be done. But the efforts invested belie, in my view, the opinion expressed by many that the military and its congressional overseers are sated and complacent men.

V

The Skelton Committee has suggested that PME must be based upon three building blocks. The first educational building block in the development of a strategist, the Committee contended, is "a firm grasp of an officer's own service, sister services, and joint commands." Second, the panel recommended that future strategists develop "a clear understanding of tactics and operational art." The examination of these two goals provides the reason, I think, that we must not confuse PME with civilian Ph.D. education. *The purpose of PME, after all is said and done, is to help officers be more efficient killers.* That is a difficult sentence to write, and nowhere is it found in the Skelton Report. But that is what, at root, PME is all about.

Senior military leaders have two fundamental operational goals: to kill (or to render impotent) the enemy or to prepare to kill him. Their PME must be directed, principally, toward that objective. That is not a pleasant thought; but it is essentially why the United States maintains military forces. That reality must never be neglected or overlooked.

The third educational building block, said the Committee, is an "understanding of the relationship between the disciplines of history, international relations, political science, and economics." The Committee believed that "[e]ach of these disciplines is critical to the formulation of strategy."[37]

Here the military has had a problem over the years. Military services have a prejudice toward the value of quantification, particularly in engineering, mathematics, and sciences. Norwich University, a military college, offered the first organized engineering curriculum in the United States. West Point, until recently, was regarded as an engineering school, the "Hudson Institute of Technology." By contrast, VMI and the Citadel have had strong liberal arts traditions. This is not a jeremiad against engineering education or the critical need in the services for technologically competent officers. I do argue, however, that the development of high competence among senior officers who are developing as strategists absolutely requires substantial knowledge in (and presumably some wisdom from) the academic disciplines cited by the Skelton Committee. Too often the undergraduate students who would major in these and allied disciplines are discouraged or denied. By the same token, hard-charging and assertive senior officers attending the war college of their service often see challenging instruction in the liberal arts at PME schools as irrelevant. That, of course, is precisely why they require that kind of education and educational renewal.

"When we can't measure the things that are important," Milton is supposed to have said, "we ascribe importance to the things we can measure."[38] If we recognize the value and vitality of technical education, as we should, we must simultaneously recognize the critical need that professional soldiers have for excellence in the liberal arts, such as philosophy. No more eloquent defender of that proposition exists than Admiral James Bond Stockdale.[39] One need not apologize for liberal education; but one can suggest that its value, in part, lies in its disciples' ability to distinguish among what should be done, what might be done, and what can be done—no easy task. If this book is correct, the leader who cannot make those distinctions will fulfill the critics' prediction: he will be incompetent, morally and mentally. Technical education helps one understand materials—and that is no small matter. Liberal education helps one understand people and morals—a literally vital matter.

This point was effectively made in testimony before the Skelton Committee by General John R. Galvin:

> Cornelius Ryan, before he wrote *The Longest Day* and his many other works of military history, was a war correspondent in Italy in World War II. I remember well his description of an experience in that war. He said he was with a couple of other corespondents as they watched a group of U.S. Army second lieutenants getting off a truck and moving up into the hills to be assigned for the first time as platoon leaders in combat. There was a silence, and one of Ryan's fellow correspondents commented, "I wonder if they are well read?" Ryan knew what this meant: these new replacements, new leaders in battle, had little experience or real preparation for what they were going to face, but maybe they at least had the knowledge of life, the philosophical outlook, the maturity, and the humanity that books can help to provide. I have often thought of that story.[40]

It is odd, in a way, that our attention is so often focused on new lieutenants, either in war or peace. Of course, that is important and serves as the rationale for the three chapters included here about my own time as a lieutenant. But we often forget that there are new colonels, too; not every commander of a wing or brigade is a seasoned

Air Force or Army 0-6. Some are entirely new to that level of responsibility. The Capstone course is intended as a means of educating (and providing some vicarious experience) to newly minted one-star generals (or 0-7's). Liberal education, then, is not a jewel in the crown. It is the crown itself.[41]

To prescribe the exact make-up of the PME curriculum is far beyond the range of this chapter and book and, I must admit, far beyond my own ken. The Skelton Committee had their building blocks right. At all levels of officer education, however, from the freshman year of the university through the PME war college level, one area of study, in particular, I think, must be emphasized: history. No one ever knows enough history. Napoleon said, or is supposed to have said, a great number of things; but one attributed comment of his that makes this point is this one:

> Peruse again and again the campaigns of Alexander, Hannibal, Caesar, Gustavus Adolphus, Turenne, Eugene and Frederick. Model yourself upon them. This is the only means of becoming a great captain, and of acquiring the secret of the art of war. Your own genius will be enlightened and improved by this study, and you will learn to reject all maxims foreign to the principles of these great commanders.[42]

The ultimate goal of PME, then, beyond ensuring command and operational competence,[43] is to provide a continuation of a broad-based liberal education.[44] The substantial liberal education, for example, will be able to help its students differentiate between relevant and irrelevant maxims "of a great but dead general" and to appreciate that wisdom is, in fact, "better than the weapons of war" (see the epigraphs to this chapter). A good education, particularly in history, political science, and international relations will also help our senior leaders to separate historical wheat from historical chaff and to comprehend which historical analogies have bearing on present predicaments.[45] In seeking to be true fighter-writers, the officers and gentlemen who have our nation's highest military trust must be educated to know both the logic of force and the force of logic. To know only the latter would make them merely pedantic; to know only the former would make them merely belligerent. Our PME system must of course impart knowledge; but beyond that the PME system must inculcate wisdom. For wisdom is the surest inoculation against the plague of incompetence.

NOTES

1. Congress, House, Committee on Armed Services, Panel on Military Education, *Professional Military Education: Hearings*. 100th Cong., 1st and 2nd sess., 9 Dec 1987–22 Sept 1988: 129. Hereafter cited as *Hearings*.

2. *The Dictionary of War Quotations*, compiled and ed. by Justin Wintle (New York: The Free Press, 1989), p. 423.

3. See, for example, Harry P. Ball, *Of Responsible Command: A History of the U.S. Army War College* (Carlisle Barracks, PA): The Alumni Association of the U.S. Army War College, 1983).

4. Discussion based upon Frederick H. Hartmann and Robert L. Wendzel, *Defending America's Security* (Washington, DC: Pergamon-Brassey's, 1988), pp. 181–182. See also pp. 183–184.

5. See Maureen Mylander, *The Generals* (New York: The Dial Press, 1974), p. 88.

6. Congress, House, Committee on Armed Services, Panel on Military Education of the One Hundredth Congress, *Report*, 101th Cong., 1st sess., 21 April 1989, Committee Print No. 4, p. 1. Hereafter cited as *Report*.

7. See, for example, Richard Hofstadter, *Anti-Intellectualism in American Life* (New York: Knopf, 1963).

8. *Dictionary* (above, n. 2), p. 438.

9. Ibid., p. 124.

10. *Report* (above, n. 6), p. 29.

11. The definition of *intellectual* is my own. But a very highly recommended reading in that regard is Roger H. Nye, *The Challenge of Command: Reading for Military Excellence* (Wayne, NJ: Avery Pub. Group, 1986).

12. *Dictionary* (above, n. 2), p. 124.

13. H. G. Wells, *The Outline of History*, II (Garden City, NY: Doubleday & Co., 1971), p. 902.

14. Mylander (above, n. 5), pp. 96–97.

15. Michael Massing, "Conventional Warfare," *The Atlantic*, January 1990: 28, 32.

16. Richard A. Gabriel, *Military Incompetence* (New York: Hill and Wang, 1985), p. 13. Gabriel and his colleague Paul Savage had earlier argued the same thesis in *Crisis in Command* (New York: Hill and Wang, 1978).

17. The term is Donald Chipman's. See his article "The Military Courtier and the Illusion of Competence," *Air University Review* 32 (March–April 1981): 53–62.

18. Gabriel, *Military Incompetence* (above, n. 16), p. 6.

19. Ibid., pp. 15, 21. Gabriel's prediction can be found in his article "The Iraqi Crisis: Sowing the Whirlwind," *International Social Science Review* 66 (Winter 1991): 13.

20. Ibid., p. 6.

21. Barry S. Strauss and Josiah Ober, *The Anatomy of Error* (New York: St. Martin's Press, 1990), pp. 246–247.

22. Ibid., p. 247.

23. Norman Dixon, *On the Psychology of Military Incompetence* (New York: Basic Books, 1976), p. 9.

24. Ibid., p. 169.
25. Ibid., p. 21.
26. Ibid., p. 168.
27. Ibid., p. 195.
28. Ibid., p. 381.
29. Eliot A. Cohen and John Gooch, *Military Misfortunes: The Anatomy of Failure in War* (New York: The Free Press, 1990), pp. 9–10.
30. Arthur Schlesinger, Jr., "Can Psychiatry Save the Republic?" *Saturday Review/World*, 7 September 1974: 14.
31. Cohen and Gooch (above, n. 29), pp. 26, 246.
32. Mylander (above, n. 5), pp. 98–99.
33. *Report* (above, n. 6), p. 165. Two reputable analysts of military affairs, Zeb B. Bradford and Frederic J. Brown, were contending as long ago as 1973 that "A profession whose stock in trade is the ability to make hard life-and-death decisions in situations of great stress must not shy away from hard decisions concerning professional education. We think that the profession would benefit from enforcement of more rigorous professional standards." In *The United States Army in Transition* (Beverly Hills, CA): Sage Publications, 1973), p. 228.
34. Discussion based upon Martin van Creveld, *The Training of Officers* (New York: The Free Press, 1990), pp. 80–95. The following two articles, although now a bit dated, present an argument different from the one I suggest in the text (about the utility of civilian graduate education for only a small selection of officers) and the one presented by van Creveld (that selected PME schools be certified to grant advanced degrees): Sam Sarkesian and William J. Taylor, Jr., "The Case for Civilian Graduate Education for Professional Officers," *Armed Forces and Society* 1 (February 1975): 251–262; and William J. Taylor and Donald F. Bletz, "A Case for Officer Graduate Education," *Journal of Political and Military Sociology* 2 (1974): 251–267. The recent book by Mark Perry, *Four Stars* (Boston, MA: Houghton Mifflin, 1989) concerns the relationship between the Joint Chiefs of Staff and our civilian leaders. Also of interest is Edward B. Glick, *Soldiers, Scholars, and the Military* (Pacific Palisades, CA: Goodyear Pub. Co., 1971).
35. "The art and science of developing and using the political, economic, and psychological powers of a nation, together with its armed forces, during peace and war, to secure national objectives." In *Report* (above, n. 6), p. 26.
36. "The art and science of employing the armed forces of a nation to secure the objectives of national policy by the application of force or the threat of force." In ibid., p. 26.
37. Ibid., pp. 30–31.
38. Quoted in Malham M. Wakin, "The Ethics of Leadership, II," in *War, Morality, and the Military Profession*, ed. by M. M. Wakin, 2nd ed., revised (Boulder, CO: Westview Press, 1986), p. 210.
39. Stockdale's fine essay from the April 1978 *Atlantic* is reprinted in ibid., pp. 10–22.
40. *Hearings* (above, n. 1), p. 1233.
41. See, for example, the fine book by J. Glenn Gray, *The Warriors: Reflections on Men in Battle* (New York: Harcourt, Brace, 1959).
42. From *Military Maxims*, trans. by Colonel D'Aguilar, 1831: #78.

43. The best reading lists I have seen are those in *The Challenge of Command* (above, n. 11).

44. On military history, see especially the essays by Jay Luvaas and B. F. Cooling, III, in *The Parameters of War*, ed. by Lloyd J. Matthews and Dale E. Brown (Washington, DC: Pergamon-Brassey's, 1987).

45. See Richard E. Neustadt and Ernest R. May, *Thinking in Time: The Uses of History for Decision-Makers* (New York: The Free Press, 1986).

8 A BRIGHT, SHINING TRUTH

> I believe that every man in uniform is a citizen first and a serviceman second, and that we must resist any attempt to isolate or separate the defenders from the defended.
>
> —Richard M. Nixon, 4 June 1969

> The nation which forgets its defenders will be itself forgotten.
>
> —Calvin Coolidge, 27 July 1920

About three-quarters of a century ago, Professor Carl L. Becker tried to summarize the "essential articles of the religion of the Enlightenment." Among those articles of faith were the beliefs that man is not natively depraved; that life's end is life here and now rather than a life after death; that reason and experience (not piety) are our best guides to perfecting our here-and-now life; and that, in order to have that good life, we have to free our minds and bodies from superstition and from arbitrary oppression by established social authorities.[1] Becker's enumeration of those elements of the Enlightenment (or the Age of Reason—about the mid-eighteenth century) is useful, for it provides something of a checklist of views and values rarely shared by the professional soldier. Post-enlightenment thinking —termed "modernity" by some philosophers—emphasizes rationalism, science, secularism, education, humanism, and belief in ineluctable progress as hallmarks of civilized order. The professional military, on the other hand, while subscribing to some extent to all of the preceding ideas, nonetheless views with great skepticism unrestrained enthusiasm or exuberance for social change. No one has improved upon Samuel Huntington's 1957 description of the worldview of the military officer. That worldview, or philosophy of life, emphasizes the "permanence, irrationality, weakness, and evil in human nature." Huntington's analysis of the military creed is so succinct that it merits extended quotation here:

> It [the military worldview] stresses the supremacy of society over the individual and the importance of order, hierarchy, and division of function. It stresses the value and continuity of history. It accepts the nation state as the highest form of political organization and recognizes the continuing likelihood of wars among nation states. It emphasizes the importance of power in international relations and warns of the dangers to state security. It holds that the security of the state depends upon the creation and maintenance of strong military forces.... It holds that war is the instrument of politics, that the military are the servants of the statesman, and that civilian control is essential to military professionalism. It exalts obedience as the highest virtue of military men. The military ethic is thus pessimistic, collectivist, historically inclined, power-oriented, nationalistic, militaristic, pacifist [in not desiring adventurous policies], and instrumentalist in its view of the military profession. It is, in brief, realistic and conservative.[2]

It is also at odds with the basic tenor and the fundamental tenets of a progressively inclined, liberally concerned, democratically structured civil order. To ignore or to deny the tension that exists between our social ethos, which is of the enlightenment, and our military ethos, which is of the Middle Ages, is to be politically autistic. To understand this concept, one need only reflect upon the idea of chivalry, which combined Christian with military ideals such as honor and courage and piety and sacrifice. The military worldview, if one accords anything of value to such ideals, is medieval in fully the best sense of the term.

One other concept ought to be explained by way of introduction to this and the following chapter. It is the idea that the existence of the military is a constant reminder to the social order of certain enduring truths more "progressive" people would prefer to forget. There is an almost startling parallel in this regard between the military ethic and the ancient discipline of political science, concerned, as at its core it must be, with "the existence of a store of objective, general truths which are as accessible to us as they were to our predecessors." If the task of traditional political science is to search for and apply those truths to the particular problems of the civil society, that discipline may well be headed for trouble, as Hans Morgenthau explains:

> For a political science which is faithful to its moral commitment of telling the truth about the political world cannot help telling society things it does not want to hear. The truth of political science is the truth about power, its manifestations, its configurations, its limitations, its implications, its laws. Yet one of the main purposes of society is to conceal these truths from its members. That concealment, that elaborate and subtle and purposeful misunderstanding of the nature of political man and of political society, is one of the cornerstones upon which all societies are founded.[3]

When the military counsels prudence and caution in a world eager for change, it courts resentment from modernists who, at times, see the rain but not the flood and feel the breeze but miss the gale. Both the American soldier (a term used generically for the representative of the military ethos) and the American social order frequently find themselves at an impasse. On the one hand, the social order would have liberties and rights and prerogatives preserved and enhanced; on the other hand, the soldier would have responsibilities and duties and obligations respected and discharged. Like Sisyphus, who was given a task destined never to be accomplished despite his most heroic efforts, the soldiers of a republic and the nations of genuine democracy must perpetually choose between those things that they must do if they are effectively to discharge their responsibilities and those things they cannot do if they are to preserve their *raison d'etre*.

For the past four decades, beginning with a number of the concerns that originated during the Korean War, the armed forces have been at the vortex of social and political change in the United States. American involvement in the Vietnam War—the longest and most divisive in American history (except, of course, for the Civil War)—aggravated the situation. American involvement in the Persian Gulf, similarly, has led to confusion and to loss of confidence, on the part of some people, in American institutions. Our society is occasionally accused of "militarism"[4] and of imperialism.[5] Historiographical debates about those subjects are, however, beyond the scope of this book. The concern of this chapter lies with what will be termed here "A Bright, Shining Truth,"[6] which is the spectacular success, since

1789 and the dawn of our Republic, with the idea of the democratic soldier. The American soldier has never been, and is not expected to be, a mindless automaton, serving unnamed masters. The very idea of a truly educated military—and the "true north" of ROTC and PME—is that the soldier can and should and must question authority, hence his orders, at certain times and places. At the same time, our armed forces, on balance, have acquitted themselves remarkably well through countless battles on land, on sea, and in the air from Bunker Hill to Baghdad. To put it very simply: this country has done something superb in creating the democratic soldier. The very idea of a democratic soldier is a bright, shining truth.

But because the term democratic soldier, to some, is an oxymoron, we must guard always against the destruction of that concept and the diminution of that reality. For the left of the spectrum wants an end to soldiery and all it represents; the left wants no reminders or the need for power, for it rarely sees utility in the employment of force. And the right of the spectrum would extend far too much privilege and prerogative to the soldiery; the right would too often enrich the soldier and impoverish the society,[7] sometimes seeing applied force as a panacea. We do ourselves, our republic, and its armed services a monumental disservice in mistaking the shadow for the substance: in any effort at military reform and restructure, if we sacrifice in any way the fundamental truth our republic has discovered (that is, the bright, shining truth of sound civil-military balance) upon the altar either of the gods of public opinion or of military extremism, we thereby destroy the very underpinnings of our constitutional order. That this is not mere prattle was underscored recently by General Colin Powell, Chairman of the Joint Chiefs of Staff:

> The U.S. Constitution is a remarkable document—and a demanding one for those of us who choose to make our career in the military. We are required to pledge our sacred honor to a document that looks at the military . . . as a necessary, but undesirable, institution useful in times of crisis; and to be watched carefully at all other times.[8]

Our country needs an army (shorthand for military service). But that army determines neither the direction nor the dénouement of our foreign policy. To be sure, our PME graduates must comprehend politics and international relations, core disciplines of national security strategy studies. On the strength of their knowledge and experience, they can guide and advise our political decision-makers; but the final power of decision is not theirs, nor should they ever claim it as within their Constitutional competencies. If we cannot maintain the balance between the two worldviews—the one of the medieval conservative soldier, the other of the enlightenment liberal citizen—we will fail to preserve the bright, shining truth of sound civil-military relations. To maintain sound civil-military relations is no easy chore. It is a true Sisyphean labor, for it is something never to be fully and finally achieved, this side of the angels.

I

Because the military has the responsibility of executing national defense policy, it is frequently, if mistakenly, attacked when it is discharging a function that may be politically unpopular. The venomous attacks on the military during the war in Vietnam are an example of this. The military has had its acerbic critics, such as Colonel David Hackworth,[9] a man indisputably rich in physical courage and undeniably poor in moral sense (and the spiritual twin of John Paul Vann); some intend their criticism destructively, and some constructively, and still others apparently to exalt only themselves. To some observers, militarism and imperialism are evident in the so-called American military-industrial complex. It is true, to some extent, that the maintenance of the defense budget (guns) is achieved not simply in terms of billions of dollars but, rather, in terms of costs in other services or commodities (butter). It is in this sense that one is reminded of Alfred Vagts's point that militarism also means "the imposition of heavy burdens on a people for military purposes, to the neglect of welfare and culture, and the waste of the nation's best manpower in unproductive army service."[10]

But it is not merely the expense of the military that rankles most. There has been the lurking suspicion—at least since the Nye Committee investigations of the 1930s—that there is an insidious partnership between the military and the merchants of death. In fact, even before the term *military-industrial complex* became a cliché, Walter Millis was pointing out that "the control of the colossal modern military machinery" was in the hands of soldiers, industrialists, and bureaucrats who exerted a "powerful influence over foreign, budgetary, and manpower policies."[11] That was the implication behind the chant "No blood for oil" yelled by protesters as the United States began bombing Iraq on 16 January 1991.

It was President Eisenhower who said, precisely thirty years before the American bombing of Iraq began, that "In the councils of government, we must guard against the acquisition of unwarranted influence, whether sought or unsought, by the military-industrial complex."[12] He exposed, thereby, a legitimate problem in American politics. The notion that the free enterprise economy is dependent upon military spending is itself, as one commentator has pointed out, a "Marxist cliché,"[13] and is the kind of obscurantist polemic that does little to help us deal with the legitimate problem diagnosed by Eisenhower. Those concerned that a conspiracy is afoot (as is Parenti: "[C]onspiracies do exist," he intones ominously[14]) are hard pressed to explain how Eisenhower, himself a career military man, somehow was overlooked for membership in the cabal. Yet even as one would be mistaken (and display paranoia) in seeing the military-industrial complex everywhere, one would be mistaken, as well, in dismissing the idea.

As dispassionate an observer of American politics as Roger Hilsman is clear about the great power exercised by military industries. And both the military and associated industries are indeed likely to have strong views on foreign policy issues. What is more, they will know how to make their voices heard effectively. Whenever any interest group in American politics can plangently state its case and powerfully influence decision-makers (especially when some of the decision-makers themselves [the military] can be seen as sitting in judgment on their own cases), there is legitimate cause for concern, if not alarm. Hilsman wisely points out, however, that "the leverage the military-industrial complex is able to exercise depends on how directly a particular issue relates to defense." Some issues are of prime military concern, such as where posts or bases may be situated in a given country overseas after the decision has been taken to have an American presence there. But on other general matters (Hilsman cites détente), the military-industrial complex has very little to say. Similarly, says Hilsman, "it is a mistake to assume that the military will always be hawkish on all foreign policy issues."[15] That is precisely the way Huntington employs his adjective *pacifist* in referring to the military's general dislike of adventurous policies—a fact often neglected in leftist analyses.

With regard to the military-industrial complex, there is a bright, shining truth that is evident, for example, in the conduct of operations in Korea, where commanders were willing, as they should have been, to sacrifice materials, even to destroy sites of cultural and religious importance, to save the lives of their troops. And that is how it should be. Our military is invariably well equipped and prepared to spend millions of dollars rather than to spend lightly the lives of American troops. Indeed, a classic example of that came in the first days of the Persian Gulf War when initial attacks on Iraqi targets were made by unmanned cruise missiles and only afterwards by coalition air attacks. Most Americans would never want to deny their military forces the logistical help they need to fight and to survive. At the same time, perhaps as episodes in the Korean and Vietnam wars demonstrate, that can be taken to extremes. While we want a well-equipped army, we do not want a flat and flabby one. The key, as always, is balance. Even as we want strong industrial contributions to our nation's security, we do not want to exalt either that security or the entrepreneurs who help to provision our troops to an extent that leads us to forget why we have defense in the first place. It is well worth remembering that what is pejoratively called the military industrial complex today was not so very long ago referred to as "the arsenal of democracy."

II

When Captain Alden Partridge founded Norwich University in 1819, he gave institutional body to his dream of the citizen-soldier. Many of the reform efforts concerned with the military over the last quarter-century have emphasized citizen, while perhaps neglecting soldier. The epigraph to this chapter by President Nixon, then speaking at the Air Force Academy, suggests, however, that it is not just radicals who wish to see servicemen as citizens first and as soldiers only second. Again, there is a tension between the two—explored in more detail in the subsequent chapter of this book.

Much of the debate over the citizen-soldier concept can be found in specific complaints about the Uniform Code of Military Justice (UCMJ), which was established by Congress in 1950 and revised in 1968 and again in 1983. The UCMJ is a set of regulations that governs the conduct of soldiers and establishes a system of courts to try soldiers accused of violating those regulations. One self-styled military critic once alleged that "Enlisted G.I.'s constitute a class that in many ways is now the most oppressed in America."[16] While that is clearly absurd hyperbole, consider some of the provisions of the UCMJ.

Under the UCMJ, for example, soldiers may be court-martialed for such offenses—some would call them innocuous—as behaving "with disrespect toward [a] superior commissioned officer" (Art. 89); feigning illness (Art. 115); using "provoking or reproachful words or gestures" (Art. 117); exhibiting "conduct unbecoming an officer and a gentleman" (Art. 133); or acting in such a way as to "bring discredit upon the armed forces" (Art. 134). Many would claim that it is surely anachronistic when a commissioned officer in the 1990s can be punished for using "contemptuous words against the President, the Vice President, Congress, the Secretary of Defense, the Secretary of a military department, the Secretary of Transportation, or the Governor or legislature of any State . . . in which he is on duty or present" (Art. 88).

The existence of such apparently reactionary articles in the UCMJ has led to some boisterous criticism by certain American reformers. One writer put it this way:

> In the eyes of professional staffers of the American Civil Liberties Union, for example, the typical court-martial is a kangaroo proceeding in which a wretched conscript [today, no doubt, a pathetic volunteer] is dragged before a panel of sadistic martinets, convicted on the basis of his own confession (which has been extracted by torture), and sentenced to fifty or sixty years of solitary confinement, chained to the wall of a subterranean dungeon, and fed on bread and water.[17]

But, in fact, as then Chief Justice of the U.S. Supreme Court Fred M. Vinson wrote in a 1953 Court opinion: "The Military courts, like the state courts, have the same responsibilities as do the federal courts to protect a person from a violation of his constitutional rights."[18] If it is true, as one writer argues, that "[w]hat the military reveres in the individual, democracy rejects; what democracy demands, the military expressly and vigorously forbids,"[19] then it may be time to recognize that Americans in uniform are, perhaps at best, nine parts citizen to one part soldier.

Almost all the questions relating to the rights of soldiers-as-civilians are rooted in interpretation of the first Amendment of the Constitution:

> Congress shall make no law respecting an establishment of religion, or prohibiting the free exercise thereof; or abridging the freedom of speech, or of the press; or the right of the people peaceably to assemble, and to petition the Government for a redress of grievances.

Even the well-known conservative columnist James J. Kilpatrick once argued that citizens do not surrender their First Amendment rights upon entering the military service.[20] Perhaps it is time to resurrect the question of unionization rights of G.I.'s, or to review the questions of collective bargaining, of free speech and political activities, and of choice of duty station, and refusal of orders on grounds of personal inconvenience or private distaste. Peter Barnes posed the question starkly: "The real question is not whether discipline will survive but whether the American military will adjust to democracy or continue to insist that democracy adjust to it."[21]

What are the soldier's rights? And where do his rights yield to his duties? One legal analyst, Joseph W. Bishop, Jr., put the answer plainly: "The [Supreme] Court has never to this day squarely held that a soldier has any constitutional rights when he is court-martialed, or indeed that he has any constitutional rights of any variety."[22] Bishop is no right-wing extremist; in fact, Mr. Justice Harry Blackmun of the Supreme Court referred to Bishop as a "noted commentator."[23] Bishop's judgment is corroborated by a long tradition of legal opinion and scholarship. In 1890, for example, Mr. Justice David Brewer wrote that "An army is not a deliberative body. It is the executive arm. Its law is that of obedience. No question can be left open as to the right to command in the officer, or the duty of obedience in the soldier."[24] As one observer put it, "These are views expressed, not by some blood-and-thunder desk-pounder, but by the Supreme Court of the United States."[25]

There does appear to be a contradiction here. Mr. Chief Justice Vinson was quoted as arguing that the military courts must preserve the serviceman's constitutional rights. And, in fact, the 1983 change to the UCMJ allowed the Supreme Court, for the first time, to review, at its discretion, certain decisions by military courts.[26] But Professor Bishop steadfastly denies that there is true precedent in the law for such preservation. The contradiction may be more of appearance than of reality. In the United States, civil and military law have a common parent—the Constitution. As the Supreme Court held in 1866:

> [T]here is no law for the government of the citizens, the armies or the navy of the United States, within American jurisdiction, which is not contained in or derived from the Constitution. And wherever our army or navy may go beyond our territorial limits, neither can go beyond the authority of the President or the legislation of Congress.[27]

The bright, shining truth of this matter is, once again, balance. The author of a standard textbook, Edward Byrne, has suggested that the role of military law is "to provide a framework for the encouragement of discipline."[28] This precise point was made by Justice William Rehnquist (now the Chief Justice) in 1974:

> This [Supreme] Court has long recognized that the military is, by necessity, a specialized society separate from civilian society.... While the members of the military are not excluded from the protection granted by the First Amendment, the different character of the military community and of the military mission requires a different application of those protections. The fundamental necessity for obedience, and the consequent necessity for imposition of discipline, may render permissible within the military that which would be constitutionally impermissible outside it.[29]

That is a precise statement of the problem—and the solution.

III

The armed forces of the United States are the most studied and carefully scrutinized military service in the world.[30] Under a microscope during both peacetime and wartime to an extent unheard of in other countries, the U.S. military also finds that it has a plethora of bosses. Although an "executive arm," as the court has held since at least 1890, the military has had its personnel policy made by the judiciary as well. As a consequence of federal court decisions, the military services have been obligated to pay turncoat POWs, to modify the manner in which they characterize administrative discharges, to eliminate compulsory chapel attendance at service academies, and to increase the compensation of some women in the armed forces.[31]

A very brief "case study review" may be helpful here to demonstrate the confusion of issues here. In 1950, when Congress adopted the UCMJ, one of its provisions was that a person who had been discharged from the military could still be tried by court-martial for certain offenses committed while in the service. In 1956, the Court held this element of the UCMJ to be unconstitutional.[32] Ex-servicemen could not be subjected to court-martial trial for crimes committed during their tours of duty. In 1957, after a woman killed her husband (an Air Force sergeant) at a base in England, the Court held that she could not be tried by the military, contending that "we reject the idea that when the United States acts against citizens abroad it can do so free of the Bill of Rights."[33] These cases served as backdrop for what becomes a very serious question in the military: what does the phrase "service-connected conduct [or crime]" mean? It is not just a bit of legal trivia.

In 1969, in the case of *O'Callahan v. Parker*[34] the Supreme Court ruled that the armed forces could not court-martial a soldier for a non-service connected crime—in this case an attempted rape that took place while the soldier was away on leave. These types of courts-martial were held to be unconstitutional because accused servicemen were denied their rights to a grand jury hearing and trial by jury, two rights guaranteed by the Fifth Amendment. Justice William O. Douglas managed to offer this animadversion against military justice: "Courts-martial as an institution are singularly inept in dealing with the nice subtleties of constitutional law." So a crime, to be within military jurisdiction, had to be service connected.

The situation was clarified a little in the case of *Relford v. Commandant*[35] when the Court unanimously agreed that the rape by a serviceman, on a military post or base, of members of another serviceman's family was a service-connected crime within the meaning of O'Callahan.[36] But questions, of course, persisted. Commanders were understandably confused about their prerogatives within the scope of the UCMJ. In a recent case, Chief Justice Rehnquist's majority opinion *overruled* the O'Callahan precedent. Rehnquist said that the service-connection test "has proved confusing and difficult for the military courts to apply." That does appear reasonable. But Mr. Justice Thurgood Marshall, dissenting, accused the majority of "contempt, both for the members of our armed forces and for its constitutional safeguards intended to protect us all." This particular case (involving a Coast Guardsman alleged to have sexually attacked children of other Coast Guardsmen off base) was kept in the military justice system.[37] (The case was *Solorio v. U.S.* [1987].)

The point of all this is that the military justice system itself is subject to intense debate. The Supreme Court itself can change—and has changed—its mind about critical concepts, such as the meaning of what is related to the service and to armed forces jurisdiction. When Captain Howard Levy refused orders to train Green Beret medics during the Vietnam War, his trial was litigated more extensively than any other military criminal case in U.S. history.[38] Levy argued that ninety percent of the prisoners in a stockade are there because of AWOL, which he regarded as a *political offense*. Levy found such things as military courtesy, rank, and military appearance to be bothersome, for he wasn't interested in "making a good impression." And if his superiors didn't like it, well, "You're going to have to take it the way it is, because that's the way it is."[39] Levy urged black soldiers to refuse to go to Vietnam because "they are discriminated against and denied their freedom in the United States," and he referred to Green Berets as "liars and thieves and killers of peasants and murderers of women and children."[40] For his refusal to obey direct orders to train medics and for violations of Articles 133 and 134 of the UCMJ—respectively, conduct unbecoming an officer and a gentleman and conduct prejudicial to "good order and discipline" in the military—Levy was charged, tried, and convicted.

Now we enter the labyrinth again. Levy's conviction was sustained by the military courts of appeal, but the U.S. Federal Court of Appeals overturned the case, contending, in essence, that Articles 133 and 134 of the UCMJ were unconstitutionally vague. The Supreme Court, in its turn, chose to review the case, and reversed the Third Circuit, holding that the articles were neither vague nor overbroad. "Just as military society has been a society apart from civilian society, so military law is a jurisprudence which exists separate and apart from the law which governs in our federal judicial establishment," wrote Justice Rehnquist. Articles 133 and 134 are, in fact, the essence of the Uniform Code of Military Justice. Howard Levy may well have been a doctor, but he was no officer. Whether he was a gentleman, one supposes, can be debated. But probably not for very long.

Military justice has had cases better known than the ones alluded to here (Calley; Slovik[41]; the Captain Jeffrey R. MacDonald case, in which this Green Beret doctor was accused of, and tried for, killing his family), but the first point to be made here is the complexity of military justice. (As William Generous has put it, "The Armed Forces are, after all, supposed to be fighting battles, not lawsuits."[42]) The second point is the need to maintain a disciplined military environment while preserving the rights we all enjoy as citizens of this Republic. There is no escaping the political nature of some cases. When President Carter reinstated draft registration early in 1980, Congress provided just enough money to register males and expressly declined to permit the registration of females. The Supreme Court eventually upheld that judgment, having to reverse a lower court's decision on the way. To use a very non-legal expression, another can of worms is the persistent question of whether servicemen can assert negligence claims against the United States. The basic concept—

known as the "Feres Doctrine"[43]—limits suits to injuries not "incident to service." Can servicemen sue for getting hurt or killed?

One would need a good law library, plenty of time, a solid legal education, and the wisdom of Solomon to sort out completely all of these kinds of issues. One must strongly agree with Chief Justice Earl Warren, however, in his view that the "courts are ill-equipped to determine the impact upon discipline that any particular intrusion upon military authority might have."[44] It was precisely in that sense that Chief Justice Vinson wrote in 1953 that

> [T]he rights of men in the armed forces must perforce be conditioned to meet certain overriding demands of discipline and duty, and the civil courts are not the agencies which must determine the precise balance to be struck in this adjustment. The Framers [of the Constitution] expressly entrusted that task to Congress.[45]

We return full circle, then, to the points made in the chapters concerned with ROTC and with PME: that the essence of military integrity lies in proper education. The military must have men and women of character, for, without them, all manner of supervision—executive and legislative and judicial—will be necessary. The stronger the character of our service personnel, the less necessary will be rules and regulations and restrictions. In 1969, Justice John Marshall Harlan dissented in the O'Callahan case. His astute observation can summarize well the arguments in this section of the book:

> The United States has a vital interest in creating and maintaining an armed force of honest, upright, and well-disciplined persons, and in preserving the reputation, morale, and integrity of the military services. Furthermore, because its personnel must, perforce, live and work in close proximity to one another, the military has the obligation to protect each of its members from the misconduct of fellow servicemen. The commission of offenses against the civil order manifests qualities of attitude and character equally destructive of military order and safety. The soldier who acts the part of Mr. Hyde while on leave is, at best, a precarious Dr. Jekyll when back on duty.[46]

Even Mr. Justice Douglas, who had reservations about the military justice system (as quoted), agreed, in principle, with Harlan:

> The Constitution gives Congress power to "make Rules for the Government and Regulation of the land and naval forces." Art. 1, [Sec.] 8, cl. 14, and it recognizes that the exigencies of military discipline require the

existence of a special system of military courts in which not all of the specific procedural protections deemed essential in Art. III [the Constitution's judicial powers article] need apply.[47]

The question of where rights end and duties begin will never be solved by those who hurl epithets or insults at one another. If a formula exists to deal with the issue, it is this: the soldier should have as much freedom as is consonant with his primary responsibility, which is to help ensure the nation's safety and security. If that is a vague formula, it is because the problem is essentially a political one, requiring the best of the political art of doing what is possible, rather than striving vainly in pursuit of what appears perfect. On balance, the judgment of F. B. Wiener seems correct:

> It should never be forgotten that if members of the armed forces were just as free to sound off as all in the civilian population indubitably are, then two predictable consequences would ensue. First, there would be no armed forces worthy of the name. Second . . . the nation would risk periodic contests between its military and civilian leaders.[48]

IV

The UCMJ is very clear that it is a court-martial offense for any soldier to disobey a lawful command, order, or regulation (Articles 90, 91, 92). At the same time, a U.S. Army field manual proclaims that:

> The fact that the law of war has been violated pursuant to an order of a superior authority, whether military or civil, does not deprive the act in question of its character of a war crime, nor does it constitute a defense in the trial of an accused individual, unless he did not know and could not reasonably have been expected to know that the act ordered was unlawful.

The same manual also states that:

> The fact that domestic law does not impose a penalty for an act which constitutes a crime under international law does not relieve the person who committed the act from responsibility under international law.[49]

It is thus entirely possible for the soldier to be caught between the competing demands of his conscience and those of his military circumstances. Since the judgments at Nuremberg and at other war crimes trials, the United States has declared itself to be squarely in favor of the soldier's right and duty of disobedience to unlawful orders.

Although the United States has no claim to purity with respect to the commission of war crimes,[50] the trial of Lieutenant William Calley, convicted for the murders of hundreds of Vietnamese around the village of My Lai, was conducted by members of his own military service and by his countrymen. No other country has done such a thing in time of war.

Because the Declaration of Independence justified the birth of the American nation as an entitlement of the "laws of Nature and of Nature's God"; because American coins and currency proclaim that "In God We Trust"; and because, since 1954, the American pledge to the flag assumes "one nation under God"—the American nation and its soldiers can hardly do other than to recognize the soldier's conscience as complementary to, and not destructive of, military discipline.

The American soldier, for his part, is urged by his Code of Conduct to "trust in [his or her] God and in the United States of America." Should circumstances require him to choose between the two, honor obliges him to be "responsible for [his] actions," and to accept willingly and manfully the consequences of that choice.

That injunction, of course, applies to officers as well as to enlisted personnel. Did it apply, too, to General Douglas MacArthur? Was MacArthur correct in defying his president in 1951? From one perspective, we must answer no. The premier concept of American civil-military relations is the idea of civilian control of the military, something MacArthur was clearly usurping in efforts to subvert the authority and orders of his commander.[51] From another perspective, MacArthur had every right to dispute his commander—once he had resigned his commission. As man and as moral agent—moreover, as soldier—MacArthur was responsible for his actions. He had to act in accordance with his informed conscience. For him to do other, or to do less, would have brought him dishonor as an officer and a gentleman. In MacArthur's case, one may well dispute both the substance of his views and the manner in which his views were expressed.

It is only the latter with which we must be concerned here. One of the bright, shining truths about the American system is our clear emphasis upon military subordination to elected, constituted civil authority. Although two authors have recently explored the bizarre "Radford Affair" (in which Navy Yeoman Charles Radford was accused of spying in 1970–1971 on the White House for the JCS in general and for Admiral Moorer in particular),[52] there is no reasonable prospect—nor has there ever been, fiction such as *Seven Days in May* to the contrary notwithstanding—for a military coup in this country. Such debate as exists about the content and direction of foreign policy is invariably the fruit of our deliberately confused system of separation of powers, designed by the Constitutional framers of 1787 to avert tyranny (even at the cost, one is obliged to concede, of governmental efficiency). The Constitution, as Professor Edward S. Corwin (1878–1963) told us, "is an invitation to struggle for the privilege of directing American foreign policy." As Justice Louis Brandeis wrote in 1926, "the doctrine of separation of powers was adopted by the Convention of 1787, not to promote efficiency but to preclude the exercise of arbitrary power. The purpose was, not to avoid friction, but, by means of the inevitable friction incident to the distribution of the governmental powers among three departments, to save the people from autocracy."[53]

No better example of separation of powers in foreign affairs exists than that of the confusion that characterized the American approach to war in 1991. President Bush insisted that he was not subject to the War Powers Resolution of 1913, which legally binds the president to observe the Constitutional stipulation that only Congress can declare war (Art. I, Sec. 8, cl. 11)—even though the President is Commander-in-Chief (Art. II, Sec. 2)—except in a state of national emergency after attack or, of course, unless Congress grants him specific statutory authorization. As the deadline of 15 January 1991 approached, at which time President Bush had indicated his willingness to use force against the Iraqi regime of Saddam Hussein, a number of Congressmen suggested that Bush had no power whatever in that regard unless empowered by Congress. Bush argued, as have all presidents since Nixon (who in 1973 vetoed the War Powers bill only to have it overriden by Congress), that the war powers bill is unconstitutional because it impinges on a clearly executive authority.[54] The Court has thus far stayed out of that fray. A constitutional crisis on the eve of U.N.-U.S. action against Iraq was averted when Bush submitted to the Congress his request for their approval of possible military action against Iraq. The Congress agreed. It is worth noting that the War Powers Resolution has been used several times, such as when Marines were sent to free the *Mayaguez* in 1975 and in 1982 when troops were sent to Lebanon. But in the Grenada rescue mission of 1983, Reagan did not deal with Congress until after troops had been dispatched; nor did the president consult Congress when troops were used in Honduras or when, in 1986, force was employed twice against Libya. At best, then, the War Powers Resolution is hortatory; that is, it encourages presidents to consult the Congress, which does have the final war-making power, at least with respect to formal declarations (a de jure power) and with respect to the power of the purse (a de facto power and a de jure power [Art. I, Sec. 8, cl. 12]).

Military personnel, with a strict chain of command, invariably find the "fudging" at the top disturbing. Used to quick and even peremptory authority, soldiers become uncomfortable in the midst of the political wrangling concerning many foreign policy and war-making decisions. This appeared to be the case, for example, with General John D. Lavelle, the commander of the U.S. Seventh Air Force in Vietnam, in 1972. He was relieved and subsequently demoted after ordering repeated and unauthorized bombing strikes of military targets in North Vietnam. Lavelle admitted that he had ordered planes to make in the neighborhood of twenty unauthorized raids on military targets in North Vietnam and had reported them to headquarters as "protective reaction" missions (meaning either an offensive or defensive response to pinpointing enemy radar controlling surface-to-air missiles or after being fired upon). The Air Force Chief of Staff, General John D. Ryan, told a congressional investigating subcommittee that at least twenty-eight unauthorized missions involving 147 aircraft had been sent out on Lavelle's orders. Offered the chance to stay on active duty as a major general (losing two stars), Lavelle demurred and was retired as a Lieutenant general (three stars). He thus became the only four-star general in modern U.S. military history to be demoted upon retirement.

A critical issue in the Lavelle case, aside from Lavelle's "interpretations [of Rules of Engagement] that were probably beyond the literal intention of the rules" (his words), was the issue of whether, and to what extent, subordinate officers and enlisted personnel had been expected to falsify reports. One Air Force junior officer, First Lieutenant D. R. Terrill, Jr., swore out court martial charges against Lavelle, contending that "any oath of allegiance requires that I at least speak out." Terrill contended that Lavelle had willfully disobeyed a direct order (about bombing sites and missions) and had lied on official reports. Another question was whether Lavelle's superiors, particular Army General Creighton Abrams, had been informed of Lavelle's bombing missions. One Air Force NCO contended that more than 200 men were involved in falsely reporting the raids. That NCO alleged that "everybody knew we were falsifying these reports. Everybody was doing it. I kept on saying 'Why' and they said, 'that's the way we do it.'" Lavelle contended that Abrams had given him permission for a number of the bombing raids that, later, had led to his relief from command. Moreover, Lavelle contended, the Chairman of the Joint Chiefs of Staff, Admiral Thomas H. Moorer, also had information about Lavelle's strikes. Abrams denied those allegations, contending, according to one Senator who took Abrams's testimony, that he (Abrams) knew about the raids but did not know they were conducted outside the ROE or Rules of Engagement. The NCO who alleged that "everybody" knew about the false reporting also contended that more than two hundred pilots of the 7th Air force at the Udoen Air Base were involved in lying on reports.

After hearing the NCO's allegations, Senator Harold Hughes of Iowa said that "the evidence has proved beyond a doubt that there was knowledge from top to bottom" in the 7th Air Force about the faked reports. Hughes was concerned about "an entire breakdown in the command structure" of that military organization. Air Force Chief of Staff General John D. Ryan told the investigating committee that he "couldn't guarantee" that such incidents "wouldn't happen again." General Ryan opposed the court-martial of Lavelle, holding that his forced retirement and demotion were adequate punishment. On 24 October 1972, the Air Force dismissed Lieutenant Terrill's complaint, contending that the service had made a "thorough investigation of all facts and material" and had dropped the charges because the "interests of discipline" had already been served.[55] On 12 October 1972, General Abrams was approved by the Senate (84–2) as Army Chief of Staff.

One reads the news reports of this time with a sense of shame and foreboding. As the Lavelle affair unfolded, the USO was acknowledging corruption in its affairs (15 June 1972). The draft was still in operation, with an announced August draft call of 8,900 (announced on 20 June). A list of twenty-four captured U.S. pilots was announced by the North Vietnamese (24 June). Acting F.B.I. Director L. Patrick Gray was announcing on 2 October that anyone who wrote that the Nixon Administration could not investigate itself was "really leveling a general indictment against all public officials." The Army was still enmeshed with the My Lai case. The records of 167 black G.I.'s ordered dishonorably discharged after a 1906 shooting of a white man in Brownsville, Texas, were ordered cleared by the Army Secretary on 28 September—the only documented case of mass punishment in the Army's history. Forty-six black and white sailors were injured on 12 and 13 October in a racial brawl on the *Kitty Hawk*. An Army private was acquitted on 14 November of killing two officers with a booby-trapped grenade at the Bienhoa air base in Vietnam in 1971—the first fragging trial of a U.S. serviceman in this country. Seven plebes were dismissed from Annapolis on 20 November because of marijuana. A former brigadier general and his secretary were indicted on 12 October of defrauding the government of thousands of dollars after a probe of Army service clubs. Secretary of Defense Melvin Laird announced on 19 October that new steps were being taken, aimed at "further strengthening" civilian control over the military.

Just to muddy the waters some more, the House Armed Services Investigating Subcommittee report released on 18 December said that the U.S. bombing strikes over North Vietnam approved in 1971 and 1972 by General Lavelle "were not only proper but essential," although the committee refrained from deciding whether the raids had been legal.[56] What in the name of all that is holy could the average American make, in 1972, of American civil-military relations? If there was confusion in the Pentagon about the Lavelle affair ("What did they know?" and "When did they know it?"—questions to crop up repeatedly during Watergate, for all that was to change was the third-person plural to the third-person singular), as well as numerous other alleged crimes and derelictions, there was confusion, too, in the Congress about whether what was proper and essential was also legal; but that confusion paled in comparison with the inchoate mess of Watergate.

If a bright, shining truth of the American system of government is our manner of handling civil-military relations, the early 1970s proved to be deeply troublesome to all patriots. It is never easy to wear the soldier's uniform, as the chapter's opening lines point out, for the soldier is an ever-present reminder of certain truths—the reality of struggle and conflict in the world—from which we normally prefer to avert our eyes. We prefer to believe the mellifluous assurances of the Pollyannas that peace is ours virtually for the asking; that all we need do is be reasonable ourselves to assure reason in others; that serenity will follow disarmament. The soldier in peacetime seems, at best, a curiosity and, at worst, an atavism, out of place and out of sentiment. But now the uniform was stained and sullied by liars, by cheaters, by thieves. And by murderers. The news from the civilian world was hardly good.[57] And it would get much worse.

V

On 1 November 1972, the Air Force Chief of Staff, General John D. Ryan, in the wake of the Lavelle Affair, disseminated a very brief policy letter to and for commanders. Excerpts follow:

> Integrity—which includes full and complete disclosure—is the keystone of military service. Integrity in reporting, for example, is the link that connects each flight crew . . . to the commander in chief . . . . Unless he is positive of the integrity of his people, a commander cannot have confidence in his forces. Without integrity, the commander in chief cannot have confidence in us. Therefore, we may not compromise our integrity—our truthfulness. To do so is not only unlawful but also degrading. False reporting is a clear example of a failure of integrity. Any order to compromise integrity is not a lawful order. Integrity is the most important responsibility of command.[58]

Ten years after Ryan's commendable statement was issued occurred yet another imbroglio, which raised the question of integrity all over again. In May 1975, former CIA analyst Samuel A. Adams reported in *Harper's* magazine that official U.S. data on VC troops had been grossly understated before the 1968 Tet offensive. In fact, real figures on VC troops were, he argued, at least twice as high as official estimates. Adams expected praise for his intelligence efforts; instead, he found himself virtually ostracized, leading, in time, to his resignation from the CIA. Adams's allegation is that General Westmoreland was able to say that "the enemy is running out of men"—a statement based, Adams says, on "fabricated numbers." What did Westmoreland know and when did he know it? Adams's efforts to lodge complaints with the CIA, with the U.S. Army, and with the Congress, he contends, were ignored. In 1982, CBS broadcast a documentary (*The Uncounted Enemy: A Viet Nam Deception*, airing on 23 January) that featured a heated exchange between General Westmoreland and CBS News correspondent Mike Wallace. In May, *TV Guide* published its own exposé, entitled "Anatomy of a Smear: How CBS 'Got' General Westmoreland." The question now became whether Westmoreland was unfairly treated by CBS and by one of their program's paid counsultants, Sam Adams. CBS admitted violations of its own canons of journalistic ethics in their program, and *Time* suggested that "Last week the CBS eye seemed to blink in embarrassment."[59]

In April 1983, the news department of CBS released an internal report that faulted its own documentary, leading to Westmoreland's suit against CBS for libel. Eleven "principal flaws"—amounting to imbalanced reporting—were discovered. CBS nonetheless stood by the "substance of the broadcast."[60] Samuel Adams himself later argued that Westmoreland finally abandoned his legal action because of certain documents in the possession of Bruce Jones, an Army intelligence lieutenant aware of the cover-up of VC numbers with which Adams had found himself embroiled. Jones's position is clear:

> In the documentary Westmoreland admitted to rejecting the intelligence estimate prepared [by two analysts] for "political reasons" and because "the people in Washington were not sophisticated enough to understand and evaluate this thing, and neither was [*sic*] the media." It was a startling admission for a professional soldier [allegedly] to make. The American military has always maintained its separation from politics, and it is, according to my 1959 ROTC manual, an "instrument of civilian authority." (The point was of course made almost two centuries before, in the Constitution of the United States.) The military does not create public or political opinion; it is controlled by it.[61]

Exactly where truth lies in this convoluted situation, one cannot be sure. Once again, both civilian and military players seem confused, if not downright untruthful. Clausewitz, of course, in *On War* had argued that "Many intelligence reports in war are contradictory; even more are false, and most are uncertain" (Ch. 6, Book I). One thing was certain on the basis of this confusion: the cause of civil-military relations in our country was not improved by this fiasco, irrespective of where final fault lay.

VI

The Iran-Contra affair is sufficiently recent and sufficiently notorious to escape a full airing here. Briefly, President Reagan had been outspoken in urging U.S. allies not to sell weapons to Iran. When an obscure Beirut magazine alleged that the U.S. government had negotiated an arms deal with Iran, supposedly a major sponsor of international terrorism, and U.S. officials finally conceded as much, it again led to confusion and outrage.

The profits from the arms sales, the government disclosed on 25 November 1986, had been secretly sent to aid the contras, fighting to overthrow the leftist government of Nicaragua. Congress (through the Boland Amendment) had proscribed aid, direct or indirect, to the contras. That was law. What did Reagan know and when did he know it? Certain members of the National Security Council were found to be deeply involved in the deal: National Security Adviser Robert McFarlane, his successor Vice Admiral John M. Poindexter, and Marine Lieutenant Colonel Oliver L. North were principals in the case.

A lengthy Congressional investigation ensued, including four weeks of joint televised hearings. As one source put it, "The hearings, plus the committee's 690-page final report, painted a picture of official deception, apparent illegality, and presidential ineptitude."[62] A retired major general (Richard Secord) and others were engaged by North to set up the "Enterprise," a private network for supporting the contras—despite the Boland Amendment.[63] North suggested that CIA Director William Casey knew about and approved the "Enterprise"; Casey died in May 1987 without being questioned in detail about his role in the illegal business.

Those watching the proceedings on television will remember Fawn Hall, North's secretary, suggesting in earnest that "sometimes you have to go above the written law." A number of documents concerned with the deals being struck were shredded. When North appeared before the investigating committee of the Congress, he wore his Marine Corps uniform, the propriety of which was questioned by a number of active duty officers, although North, technically, violated no regulation by so doing. "I will tell you right now, counsel, and all the members here gathered that I misled the Congress [in earlier testimony]."[64] Why was that? Why had he lied? "Congress is to blame because of the fickle, vacillating, unpredictable, on-again, off-again policy" toward the contras.[65] This from an O-5 in the Marine Corps; this from a lieutenant colonel elected to no position of authority in our democratic government; this from an officer sworn to uphold and defend the Constitution of the United States; this from an Annapolis graduate for whom the honor code had supposedly become a way of life; this from one man who had superciliously arrogated to himself the province of American foreign policy.[66]

The upshot, by the way, was an "Olliemania" in our Republic. Few doubt the patriotism and good intentions of Oliver North. That he became a new American hero is understandable; it is also too bad. Senator Daniel K. Inouye, also a patriot and heroic war veteran, contrasted North with other American heroes who were acclaimed for their honesty, calling it "painful to learn from your testimony that you lied and misled because of what you believed to be a just cause—support of Nicaragua's freedom fighters." Senator Inouye said that North, in opposing communism, had adopted "one of the most important tenets of communism and Marxism: the ends justify the means."[67]

A very different case—perhaps too much truthfulness—came to light in late 1990. Air Force Chief of Staff Michael J. Dugan was fired on 17 September for publicly disclosing confidential elements of U.S. military strategy in the Persian Gulf. Dugan became the first member of the Joint Chiefs of Staff to be fired since Admiral Louis E. Denfield, a Chief of Naval Operations, in 1949. In fact, Dugan was the highest-ranking officer to be relieved of duty since MacArthur in 1951. Certain of Dugan's remarks—such as his denigration of Iraqi forces, his boasting about U.S. air power and his suggestion that other U.S. forces would have only a subsidiary role in the war, and his revelation of the numbers of some U.S. forces— led to the decision of Secretary of Defense Cheney to fire Dugan. According to one news source, "Dugan's candor infuriated the Bush administration."[68] National Security Adviser Brent Skowcroft even found it necessary to explain that the Air Force Chief of Staff was not in the chain of command and was not speaking for the administration. Dugan clearly deserved a reprimand, but one must wonder whether, particularly in the light of subsequent events, he deserved to be summarily relieved.

VII

A major objective of all security policy is to ensure the survival of the nation. But survival alone makes no sense unless it is in the service of some fundamental values that imbue the nation's existence with some meaning and purpose. The late Dean Acheson, a former Secretary of State, put it well: "[T]he means we choose to overcome the obstacles in our path must be consonant with our deepest moral sense." When the CIA allegedly was considering the murder of Fidel Castro, Senator Frank Church said: "The notion that we must mimic the communists and abandon our principles [is] an abomination."[69]

Exceptions, ironically, can help prove rules. The horror of the My Lai massacre and the criminal conduct obvious in some of the cases alluded to above are causes of confusion, of anger, and of resolve that they will not be repeated. The bright, shining truth of our Republic is the recognition that, while the existence of power must be understood and respected by those who are charged with the preservation of our security, deference to integrity and to truth and to justice must always supersede the security impulse.

The practical ramifications of this feeling about the vitality of morality in statecraft go to the root of American military and diplomatic principles. There have been, it must be admitted, a number of flagrant examples of departure from this theme. But one must never forget that crimes committed, for example, by American Army lieutenants, are in complete and clear contradiction to the policies they were expected to uphold; and that lies told for example, by Marine lieutenant colonels in support of policies hatched in contradiction to the laws and institutions to which they are expected to maintain professional fidelity violate the very code of integrity the military is charged with preserving.

"Who would dare explain American policy as derived from our 'sacred honor'?" asked Max Ways. "The phrase has become worse than hollowly archaic; moderns find it downright offensive." Such a phrase, of course, may be found in our Declaration of Independence, but not in the contemporary lexicon of political science. "Yet it is with 'sacred honor' that the political function of the people has most to do." He adds: "Freedom is what we need so that we may bear our moral responsibility for what the government does in our names."[70]

An aspect of national security that must not be overlooked is the problem of preserving our internal values from *internal threats*. That astute observer of American society, Alexis de Tocqueville, wrote that "Nothing is so dangerous as an army amidst an unwarlike nation; the excessive love of the whole community for quiet continually puts the constitution at the mercy of the soldiery." He went on to argue that "The general spirit of the nation being infused into the spirit peculiar to the army, tempers the opinions and desires engendered by military life, or represses them by the mighty force of public opinion.

Teach but the citizens to be educated, orderly, firm, and free, and the soldiers will be disciplined and obedient."[71] Writing in *The Federalist*, a series of 85 essays composed in 1787 and 1788 by Hamilton, Madison, and Jay to urge ratification of the proposed American constitution, Alexander Hamilton expressed a judgment similar to Tocqueville's:

> Safety from external danger is the most powerful director of national conduct. Even the ardent love of liberty will, after a time, give way to its dictates. The violent destruction of life and property incident to war, the continual effort and alarm attendant on a state of continual danger, will compel nations the most attached to liberty to resort for repose and security to institutions which have a tendency to destroy their civil and political rights. To be more safe, they at length run the risk of being less free.[72]

In a similar sense, Benjamin Franklin once had Poor Richard admonish, "They that can give up essential liberty to obtain a little temporary safety deserve neither liberty nor safety."

Writing more than 2,000 years ago, Thucydides, in the first book of his classic work, *The Peloponnesian War*, has Pericles utter a timeless warning: "I fear our own mistakes far more than the strategy of our enemies." To resolve all the complexities of the status of the citizen-soldier is truly a Sisyphean task, for one supposes that it will not finally be completed. Nor will all the answers be clear, at every juncture, in the course of American foreign policy. Our nation's soldiers and our nation's leaders must chart a prudent course between the Scylla of naive utopianism and the Charybdis of unprincipled Machiavellianism. No simple choices exist.

Individual and national military choices are part of the human condition. Like Sisyphus, American soldiers and statesmen are condemned by the circumstances of an imperfect world to try to solve problems that admit no full and final solutions. The truth of our Republic is as bright and as shining as is the persistence and the nobility of our efforts. Perhaps, with Albert Camus, we may find some consolation in the knowledge that "The struggle itself toward the heights is enough to fill a man's heart. One must imagine Sisyphus happy."[73] The democratic soldier, like the government he serves, is a magnificent anomaly in world history. That our Republic takes a soldier's rights and his duties seriously and that our Republic's finest leaders are animated by a sense of honor—that is our bright, shining truth.

NOTES

1. Carl L. Becker, *The Heavenly City of the Eighteenth-Century Philosophers* (New Haven, CT: Yale University Press, 1932), pp. 102–103.

2. Samuel P. Huntington, *The Soldier and the State* (Cambridge, MA: Belknap Press, 1957), p. 79.

3. Hans J. Morgenthau, "The Nature and Limits of a Theory of International Relations," in *Theoretical Aspects of International Relations*, ed. by William T.R. Fox (Notre Dame, IN: University of Notre Dame Press, 1959), pp. 19, 22.

4. Laurence I. Radway, "Militarism," *International Encyclopedia of the Social Sciences*, 1968, X: 300.

5. Michael Parenti, *The Sword and the Dollar* (New York: St. Martin's Press, 1989). See, especially, chapters 2, 4, and 5. For more of the same, see C. Wright Mills, *The Causes of World War Three* (New York: Simon and Schuster, 1958) or his earlier work, *The Power Elite* (New York: Oxford University Press, 1956).

6. A play, of course, on Neil Sheehan's *A Bright Shining Lie* (New York: Random House, 1988), a biography of John Paul Vann.

7. President Reagan's defense build-up cost about $2 *trillion*. Defense Secretary Cheney was saying in 1991 that there would be no more blank checks, and he canceled the Navy's $57 billion attack bomber, the A-12. See *Time*, 21 January 1991: 45.

8. *U.S. News & World Report*, 4 February 1991: 27.

9. David H. Hackworth and Julie Sherman, *About Face* (New York: Simon and Schuster, 1989). This is *must* reading for the professional. "Overall," says William L. Hauser, "this is perhaps the most important book yet to come out of the Vietnam War. It will be dismissed by those who do not wish to learn the ugly but salutary lessons of that conflict but who will justify that decision by citing an ad hominem reason, the eccentricities and alleged misdeeds of the author. That will be a shame, for truly—*he is right.*" In *Armed Forces & Society* 16 (Winter 1990): 315. See also the review of Hackworth's book in *Army* (June 1989): 83–87 and the responses in August (pp. 5–7)

10. Alfred Vagts, *A History of Militarism*, Greenwich ed. (n.p.: Meridian Books, Inc., 1959), p. 14.

11. Walter Millis, *Arms and Men* (New York: Capricorn Books, 1956), p. 346.

12. Quoted in Stephen E. Ambrose and James A. Barber, Jr., eds., *The Military and American Society: Essays and Readings* (New York: Free Press, 1972), p. 63.

13. Adam Yarmolinsky, *The Military Establishment: Its Impacts on American Society,* Perennial Library, abridged ed. (New York: Harper & Row, 1971), p. 235.

14. Parenti (above, n. 5), p. 195.

15. Roger Hilsman, *The Politics of Policy Making in Defense and Foreign Affairs*, 2nd ed. (Englewood Cliffs, NJ: Prentice Hall, 1990), pp. 309–311. See also Bob Woodward, *The Commanders* (New York: Simon & Schuster, 1991), p. 314.

16. Peter Barnes, *Pawns: The Plight of the Citizen-Soldier* (New York: Warner, 1971), p. 26. For more of the same, see Robert S. Rivkin, *G.I. Rights and Army Justice* (New York: Grove Press, Inc., 1970); and Robert Sherrill, *Military Justice is to Justice as Military Music is to Music* (New York: Harper & Row, 1970). In

"Against the Evidence," *Commentary* (June 1971): 92. Professor Bishop (see also below, n. 17) says of Sherrill's book that "The listing and documentation of Mr. Sherrill's errors, great and small, could literally be prolonged until the review is longer than the book."

17. Joseph W. Bishop, Jr., *Justice Under Fire* (New York: Charterhouse, 1974), pp. 19–20.

18. *Burns v. Wilson* 346 U.S. 137, 142 (1953).

19. Peter G. Bourne, "The Military and the Individual," in *Conscience and Command*, ed. by James Finn (New York: Vintage Books, 1971), p. 153.

20. James J. Kilpatrick, syndicated column, *The South Bend Tribune*, 14 March 1977: 12.

21. Barnes (above, n. 16), p. 239. Since 1978, membership in a union organization intending to alter the terms or conditions of military service has been specifically prohibited. Public Law 95-610 (8 November 1978).

22. Bishop (above, n. 17), p. 114.

23. *Parker v. Levy* 417 U.S. 733, 763 (1974).

24. *In re Grimley* 137 U.S. 147, 153 (1890).

25. Frederick B. Wiener, "Are the General Military Articles Unconstitutionally Vague?" *American Bar Association Journal* 54 (April 1968): 362. Wiener argued in the negative in this 1968 article. Six years later, the U.S. Supreme Court agreed with him, as the chapter text will explain. See *Parker v. Levy* 417 U.S. 733 (1974); and *Secretary of the Navy v. Avrech* 418 U.S. 676 (1974). For the briefs of counsel to research these cases, see, respectively, 41 *U.S. Supreme Court Reports*, 1245-46 and 1316-17. Wiener also wrote "The Perils of Tinkering with Military Justice," *Army* (November 1970): 22–25.

26. *Facts on File*, 9 December 1983, p. 929.

27. *Ex parte Milligan* 71 U.S. 2, 141 (1866).

28. Quoted in Philip Gold, "Military Law: Limited, Not Inferior," *Insight*, 13 April 1987: 24.

29. *Parker v. Levy* 417 U.S. 733, 743, 758 (1974).

30. See, for example, Franklin D. Margiotta, ed., *The Changing World of the American Military* (Boulder, CO: Westview Press, 1978).

31. Michael F. Noone, Jr., "Military Social Science Research and the Law," *Armed Forces and Society* 15 (Winter 1989): 193.

32. *Toth v. Quarles* 350 U.S. 11 (1955).

33. *Reid v. Covert* 354 U.S. 1 (1957). A useful guidebook for the legal layman is Charles Shanor and Timothy P. Terrell, *Military Law in a Nutshell* (St. Paul, MN: West Publishing, 1980.

34. *O'Callahan v. Parker* 395 U.S. 258 (1969).

35. *Relford v. Commandant* 401 U.S. 355 (1971).

36. For more background, see William T. Generous, Jr., *Swords and Scales* (Port Washington, NY: Kennikat Press, 1973).

37. *Facts on File*, 1987, p. 460.

38. *Parker v. Levy* 417 U.S. 733 (1974). See Norman Thorpe, "Discipline and Justice in the Armed Forces," *Air University Review* 33 (Nov–Dec 1981): 64.

39. Quoted in Finn (above, n. 19), pp. 168, 184. The case may be found at *Parker v. Levy* 417 U.S. 733 (1974).

40 Quoted in Shanor and Terrell (above, n. 33), p. 159; see also *Facts on File*, 1973, p. 313, and 1974, p. 501.

41. See William Bradford Huie, *The Execution of Private Slovik* (New York: Signet, 1954).

42. Generous (above, n. 36), p. 204.

43. *Feres v. United States* 343 U.S. 135 (1950).

44. Quoted in Noone (above, n. 31), p. 202.

45. *Burns v. Wilson* 346 U.S. 137, 140 (1953). Cf. the opinion in *Orloff v. Willoughby* 345 U.S. 83, 94 (1953): "The military constitutes a specialized community governed by a separate discipline from that of the civilian."

46. *O'Callahan v. Parker* 395 U.S. 258, 281 (1969).

47. Ibid., p. 261.

48. Wiener (above, n. 25), p. 363.

49. U.S. Department of the Army, *The Law of Land Warfare*, Field Manual 27-10 (Washington, DC: Government Printing Office, 1956), pp. 182–183.

50. See Clifton D. Bryant, *Khaki Collar Crime* (New York: Free Press, 1979).

51. John Spanier, *The Truman-MacArthur Controversy and the Korean War* (Cambridge, MA: Belknap Press, 1959). See also chapter 7 of Walter Millis with Harvey C. Mansfield and Harold Stein, *Arms and the State* (New York: Twentieth Century Fund, 1958).

52. See the controversial book by Len Colodny and Robert Gettlin, *Silent Coup: The Removal of a President* (New York: St. Martin's Press, 1991), Chs. 1–4 and pp. 382 and 391.

53. Quoted in Louis Henkin, "Foreign Affairs and the Constitution," *Foreign Affairs* 66 (Winter 1987–1988): 299n. See also Henkin's superb book of the same title (Mineola, NY: Foundation Press, 1972).

54. For a different perspective to the War Powers Resolution, see Francis D. Wormuth and Edwin B. Firmage, *To Chain the Dog of War*, 2nd ed. (Urbana, IL: University of Illinois Press, 1989).

55. *Facts on File*, 1972, pp. 440–442, 465, 499, 745, 783–784, 807–808, 956.

56. *Facts on File*, 1972., p. 1049.

57. For example, see David Halberstam, *The Best and the Brightest* (New York: Random House, 1972).

58. Quoted in Malham M. Wakin, ed., *War, Morality, and the Military Profession*, 2nd ed., revised (Boulder, CO: Westview Press, 1986), p. 180.

59. *Time*, 26 July 1982: 40.

60. *Facts on File,* 1983, pp. 406–407.

61. Bruce E. Jones, *War Without Windows* (New York: Berkley, 1987), p. 265.

62. Donald Morrison, "The Iran-Contra Affair," *1988 Brittanica Book of the Year* (Chicago, IL: Encyclopedia Brittanica, Inc., 1988), p. 485.

63. Ibid.

64. Congressional Quarterly, Inc., *The Iran-Contra Puzzle* (Washington, DC: Congressional Quarterly, Inc., 1987), p. 121.

65. Ibid.

66. See William S. Cohen and George J. Mitchell, *Men of Zeal* (New York: Viking, 1988).

67. *Facts on File*, 1987, p. 511. See also *The Iran-Contra Arms Scandal: Foreign Policy Disaster* (New York: Facts on File Publications, 1988).

68. *Facts on File*, 1990, pp. 688–689. See also chapter 20 in Woodward (above, n. 15).

69. Acheson is quoted by William H. Vatcher, Jr., *Panmunjom: The Story of the Korean Military Armistice Negotiations* (New York: Praeger, 1958), p. 220. Church is quoted in *Time*, 16 June 1975: 10.

70. Max Ways, *Beyond Survival* (New York: Harper and Brothers, 1959), pp. 5, 9.

71. Alexis de Tocqueville, *Democracy in America*, ed. and abridged by Andrew Hacker, trans. by Henry Reeve (New York: Washington Square Press, 1964), pp. 284–285, 287.

72. Alexander Hamilton, James Madison, John Jay, *The Federalist Papers*, 8 (1787).

73. Albert Camus, *The Myth of Sisyphus and Other Essays*, trans. Justin O'Brien (New York: Knopf, 1961), p. 123.

9 RHAPSODY IN BLUE: THE AMERICAN MILITARY ETHIC

> What a society gets in its armed services is exactly what it asks for, no more and no less. What it asks for tends to be a reflection of what it is. When a country looks at its fighting forces, it is looking in a mirror; if the mirror is a true one, the face that it sees will be its own.[1]
>
> —Sir John W. Hackett

> Those who say life is worth living at any cost have already written for themselves an epitaph of infamy, for there is no cause and no person they will not betray to stay alive.[2]
>
> —Sidney Hook

This chapter is a paean—a rhapsody—about something we in the United States do, on balance, uncommonly well. In the course of more than 200 years of history since George Washington became our first commander-in-chief, we have learned a number of lessons about democratic government, one of which is that while military forces are necessary in a dangerous world, those same forces pose a threat to the very institutions whose security they are charged with providing. That we need to be reminded of this doleful fact is itself testimony to the "bright, shining truth" discussed in the preceding chapter. Among the substantial and enduring political achievements of our democratic republic is the development of the American military ethic.

All the American armed forces wear blue in at least one of their uniforms. According to the Research Division of the USAF Historical Research Center at Maxwell AFB, Alabama, the color blue in Air Force heraldry means such things as *constancy, devotion, loyalty, piety,* and *truth.* That is the crux of the American military ethic (AME). It is in precisely this sense that Sir John W. Hackett addressed cadets at the Air Force Academy more than twenty years ago: "[T]he highest service of the military to the state may well lie in the moral sphere."[3]

It is, to be sure, a paradox of the first order that a profession charged with killing can—indeed must—be entrusted as well with inspiring and exalting those values that impart meaning and purpose to national life. Clearly, the military is not—and must never be—the sole repository of national resolve. Max De Pree had it exactly right in saying that "Leaders must take a role in developing, expressing, and defending civility and values."[4] By practically all accounts, military commanders are leaders in the full sense that De Pree suggests; they therefore have ethical, as well as soldierly, responsibility. Their job is to defend American core values by moral as well as by physical means.

But, one must inquire, how can one defend values unless he is able to determine which values are defensible? If leaders help to develop values, aren't values initially instrumental in developing leaders? Thus, we return again to the idea of proper liberal education, without which (at both the ROTC, or undergraduate, level and the PME level) there can be no leaders, hence no heritage of values to be preserved and promulgated, hence no military ethic, hence no "rhapsody in blue." As Admiral Stockdale put it:

> One who leads men into battle or organizations into booby-trapped territory, while committed to the idea that each empirically unverifiable value judgment is just as good as the next, is in for trouble. . . . Thus, in addition to history, I recommend concentrating on philosophy and high quality "ultimate situation" literature. By that I mean selections from the Book of Job, the Socratic dialogues of Plato, the ethical writings of Aristotle, Epictetus, Kant, and Mill, as well as readings from Emerson, Dostoevsky, Melville, Conrad, Koestler, Monod, and Solzhenitsyn. This somewhat intimidating fare can be served up by a teacher who has himself led people under pressure in a way that illuminates the basic principles of right, good, honor, duty, freedom, necessity, law, and justice as they apply to the human predicament generally and to the role of the leader in particular.[5]

If we consider values to be just ethics (that is, the ethics of justice), then education and leadership are not only compatible; they are symbiotic. Politics, then, is not merely the study of power: it is the study, as Russell Kirk once told us, of "the application of ethics to the concerns of the commonwealth." Do we have a moral military? We probably then have a moral society. Do we have a debauched military? We probably then have a debauched society.

As one professional military officer wrote two decades ago: "There is probably a good deal of truth in the adage that a democracy gets the military establishment it deserves."[6] Denis Brogan, the Tocqueville of our age, agreed: "A country has the kind of army its total ethos, its institutions, resources, habits of peaceful life, make possible to it."[7] Thus, when we discuss the AME, we inquire into the very nature of our social order. As with almost any inquiry into practically any serious subject, we will find there reason for shame and pride; for sorrow and joy; for change and preservation.

I

There are two models or paradigms of the American military ethic. As is true of virtually all models, these efforts at representing reality are rather too simplistic. Still, the models offer a starting point for an inquiry into the nature of the soldier's creed or, to use the German word, *Weltanschauung* (worldview). One model is called the "traditionalist," or the "separatist," or the "professional soldier" model. This suggests that the soldier is invariably apolitical, uninvolved—resolutely so—in politics. The soldier fights and prepares to fight; by calculation, he has nothing to do with politics; by the same token, there is, from this vista, no intimacy between war and diplomacy. The two are distinctly segregated. Listen, now, to MacArthur:

> Let civilian voices [he told the West Point Corps of Cadets in his stirring 1962 address] argue the merits or demerits of our processes of government: Whether our strength is being sapped by deficit financing indulged in too long, by Federal paternalism grown too mighty, by power groups grown too arrogant, by politics grown too corrupt, by crime grown too rampant, by morals grown too low, by taxes grown too high; whether our personal liberties are as thorough and complete as they should be.
>
> These great national problems are not for your professional participation or military solution. Your guidepost stands out like a tenfold beacon in the night: duty, honor, country.[8]

Of course, MacArthur had rarely taken his own advice during his distinguished career! In fact, MacArthur personifies the modernist approach, although the general would have hotly denied it. "During World War II," historian Samuel Eliot Morison explained, "he had largely confined his extreme [political] views to conversations, but now [during the Korean War] he felt strong enough to appeal over the President's head to the Republican party and public opinion. He never crossed the Rubicon, to be sure, but his horse's front hoofs were in the water."[9]

Another model may be called the "modernist," or "fusionist," or "soldier-statesman" model. One aspect of this approach, as Samuel Huntington pointed out a quarter-century ago, was that "military leaders incorporate political, economic, and social factors into their thinking."[10] This model, which is, of course, the point and purpose of the Skelton Committee's suggested reforms concerning the development of a corps of national strategists who are capable of integrating military analysis with political and diplomatic appraisal, is clearly the more common view of the military since the end of World War II. The great commanders of World War II—King and Marshall, Arnold and Eisenhower—helped to initiate the movement. Certainly they had to. But this is not to suggest, however, that they always grasped the geopolitical ramifications of their decisions, such as Eisenhower's and Marshall's reluctance to seize Prague and Berlin toward the end of the war.[11] Ironically, the most effective soldier-statesman may well have been—MacArthur. Morison quotes the general as saying: "Could I have but a line a century hence crediting a contribution to the advance of peace, I would yield every honor which has been accorded by war." Responds Morison:

> Here's your line, General; this historian salutes you. Your efforts for peace and good will [in postwar Japan] entitle you to a place among the immortals. No proconsul, no conqueror in ancient or modern times succeeded to the degree that you did in winning the hearts of a proud and warlike people who had suffered defeat. Your victory was a dual one—military, and in the highest sense, spiritual.[12]

Thus, MacArthur paid lip service to the idea of the separatist model, but, in fact, he typified—for better and for worse—the fusionist model.

During and after the Korean War, the fusionist approach became increasingly acceptable—in part because of MacArthur's "fading away." There had always been the lurking suspicion that General

MacArthur, even while on active duty, had wanted one more laurel, the presidency. But no one but the lunatic fringe feared George C. Marshall as a would-be "man on horseback." Eisenhower, a five-star general, became president—not a uniformed dictator. And Kennedy, as Jerome Slater put it, actually ordered the military, "from the joint chiefs right down to academy cadets to eschew 'narrow' definitions of military competence and responsibilities, [to] take into account political considerations in their military recommendations, and [to] prepare themselves to play active roles in the policy-making process."[13] Slater contends that the triumph of this model should be a matter of concern to Americans, because the military, he argues, is, well, too military![14]

The prospect of returning, as it were, to a nineteenth-century U.S. military establishment, wholly separated from political discourse, is neither practicable nor desirable. One need read only Clausewitz to understand the reasons for that. Twenty-five centuries ago, in fact, Sun Tzu pointed out that "The art of war is of vital importance to the state. It is a matter of life and death, a road either to safety or to ruin. Hence under no circumstances can it be neglected."[15] By this logic, then, the symbiotic nature of power and politics is clear. Not only must our civilian policy-makers understand the employment of power, but our military leaders must understand the art of politics.

In an excellent essay, John Garrison suggests that the traditional American concept of civilian control of the military is beyond dispute and that if by *politics* we mean the rough-and-tumble world of partisan challenge, then, of course the military must be apolitical. "The military profession and its professionals in America," he points out, "have remained remarkably free from partisan party politics."[16] But if we understand politics as the process by which groups compete for, say, allocations, then the military, practically by necessity, is a political actor—and there is no purpose served in denying it. And there is nothing nefarious about it. Indeed, as Garrison points out, our very system of separation of powers contributes to military involvement in politics:

> Since virtually all such alternatives in the national security policy field involve some aspect of military force, it follows that experts in the development and application of such force—military professionals—*must* be involved in the decision-making process.[17]

The modern view of the soldier conversant in diplomacy and

politics in no way, then, augurs a military take-over! Quite the reverse seems true. In 1962, during the Cuban missile crisis, as reported by Robert Kennedy:

> President Kennedy was disturbed by this inability [of some of his military advisers] to look beyond the limited military field. When we talked about this later, he said we had to remember that they were trained to fight and to wage war—that was their life. Perhaps we would feel even more concerned if they were always opposed to using arms or military means—for if they would not be willing, who would be? But this experience pointed out for us all the importance of civilian direction and control and the importance of raising probing questions to military recommendations.[18]

Senior military officers ignorant of the possible or probable diplomatic or military consequences of their recommendations are every bit as dangerous, Kennedy suggests, as military officers inept in their own craft. The model of the soldier innocent of politics, pristine in the face of electoral battle, can no longer be given credence; at the same time, however, the opposite image of some soldier-despot seizing power in the Republic is equally nonsensical. When Arthur Schlesinger, Jr. wrote about the "imperial" inclination in this country in the late 1960s and into the mid-1970s, he was discussing civilian presidents. And as John Lovell has written:

> If the Vietnam experience has taught us anything, it should have revealed that "the best and the brightest" products of Harvard, MIT, and other prestigious civilian institutions, many of these individuals with advanced academic or professional degrees, were capable of monumental errors of judgment.[19]

Precisely to the extent that our military leaders are educated professionals, the problem of the modernist model, or the fusionist model, or the soldier-statesman model is no problem at all.

II

The suggestion of a separatist ideal for the soldier (no political involvement) as opposed to the fusionist ideal (political involvement) in fact begs the central question: what kind of political involvement will there be? If the involvement is like that of MacArthur in Japan, it is desirable and consonant with the American military ethic. If the involvement is like that of MacArthur in Korea, it is undesirable and contradictory to the American military ethic. A parallel can be drawn to ethics. On the one hand, no one wants an army of ethical cretins, incapable of moral judgment. Indeed, since the war crimes trials at the end of World War II, the old notion of *Respondeat Superior* (let the superior officer respond for the purported crimes of his underling) is no longer recognized in and by international law. Soldiers *must* make prudential value judgments; they *must* disobey illegal orders. Consequently, if they are "ethical cretins," they cannot discharge their *military* duties. Similarly, if soldiers are ethical cataleptics, they will have a kind of abulia—an inability to choose or to decide. By "ethical catalepsy," I refer to what some people call "paralysis by analysis"—a morbid fear of doing anything, lest something bad be done. In this sense, Lloyd Matthews wrote recently about the "Moralitymonger," who "can't take a step without encountering an ethical affront."

> He sets himself up as the conscience of the Army and its members, and is forever on guard against ethical contamination. His nose is anchored to a 180 degree swivel so that it can sniff out sin right and left; his eyeballs are on long, flexible stalks so that no wickedness before or behind escapes his searching gaze; and his ears are marvelously attuned to the least whisper of scandal, real or imagined, from all points of the compass.[20]

Clearly, we must have a politically fusionist ethic if we are to have soldiers with diplomatic sense and sensitivities. Equally clearly, we must have conscientious soldiers—but not to the point of being officious or overscrupulous. An intelligent middle ground is afforded to us by understanding what a profession is all about. Huntington's definition is the standard: the professional has expertise, responsibility, and corporateness (a shared "sense of organic unity and consciousness of themselves as a group apart from laymen").[21] Webster's *Third New International Dictionary* provides this view of *profession*: "the act of taking the vows that consecrate oneself to special religious service." Or "an act of openly declaring or publicly claiming a belief, faith, or opinion." To the extent that our military is truly professional, it will be knowledgeable; it will accept its accountability to the society it serves; it will prove loyal to its own long-standing canons (a word I am not trying to pun here!). Those canons, of course, are the ethical imperatives of the AME.

In order to examine the AME, one is best advised to work backwards—that is, to explore *ethic*, then *military*, and finally *American*.

An *ethic*, says the 1901 Oxford Dictionary, is "[t]he rules of conduct recognized in certain associations or departments of human life." By the same token, Webster's *Third New International Dictionary* defines *ethos* as "the complex of fundamental values that underlies, permeates, or actuates major patterns of thought and behavior in any particular culture, society, or institution." The difficulty lies in recognizing that practically any group, however malevolent, can have an ethos. Traditionally, the military ethos or ethic or philosophy or worldview has been associated with the ideal of *honor*, defined by that 1901 Oxford Dictionary as "a fine sense of and strict allegiance to what is due or right." In tying the ethos of the military to honor, which renowned sociologist Morris Janowitz invariably did ("'Honor' is the basis," he wrote, "of [the military's] belief system."[22]), one retreats almost immediately to the idea of the professional—"taking vows," as defined above, "that consecrate oneself to special religious service." Analyst Anthony Hartle contends that "a professional ethic will generally perform . . . three critical functions . . . : it protects society from exploitation, it enhances the image of the professional, and it may articulate a warrant for certain actions morally impermissible for a nonprofessional."[23] The ethic of the military, in a very real and palpable sense, must be "religious" or "sacred."

There is no finer expression of this ideal than MacArthur's, from a 1935 speech:

> The military code which he [the American soldier] perpetuates has come down to us from even before the age of knighthood and chivalry. It embraces the highest moral laws and will stand the test of any ethics or philosophies ever promulgated for the uplift of mankind. Its requirements are for the things that are right, and its restraints are from the things that are wrong. Its observance will uplift everyone who comes under its influence. The soldier, above all other men, is required to perform the highest act of religious teaching—sacrifice. In battle and in the face of danger and death he discloses those divine attributes which his Maker gave when he created man in his own image. No physical courage and no brute instincts can take the place of the divine annunciation and spiritual uplift which alone sustain him. However horrible the incidents of war may be, the soldier who is called upon to offer and to give his life for his country is the noblest development of mankind.[24]

MacArthur sounds strikingly like Cardinal Newman, the advocate of truly liberal education, whom we quoted in chapter 6. The cynic can complain that MacArthur's comments are maudlin and mawkish. But one could as easily object to the "sentimental" adjuration: "Treat others the way you would have them treat you" (Mt. 7:12). The naval strategist Alfred Thayer Mahan suggested in 1897 that, above all, the officer must have a sense of the moral to perform his duties:

> Let me try to state clearly what qualities, in my opinion, you should especially wish in the line officer of the Navy. I scarcely think I can err in assigning to the foremost place moral power; strength to control self and others; fearlessness in responsibility and danger; self-reliance; promptitude in action; readiness of resource; calmness amid excitement. No amount of mental calibre, far less any mere knowledge, can compensate for a deficiency in moral force in our profession.[25]

An *ethic*, of course, need not be ethical; but the military's ethic must be ethical, for a military ethos devoid of integrity leads to perversion on a scale possibly so massive that it practically defies description. That is the reason why, when we discuss the ethos of the military, we must, as MacArthur and Mahan did, simultaneously discuss moral power. Colonel Hartle has suggested that the military ethos is shaped by the "functional requirements of military service"—something that the Supreme Court has thus far recognized in its deliberations about the UCMJ as we explored in the preceding chapter; by the international laws of war—which, since the war crimes trials at Nuremberg and elsewhere after World War II, recognize the vitality of natural law and of "right reason" in insisting that illegal orders must be disobeyed; and by "the core values of American society."[26] The key is to examine the term *military*.

III

Harold Lasswell's 1941 observation that soldiers are "specialists on violence" is the standard by which the armed forces are judged.[27] An intra-military debate has raged for years over the question of whether Lasswell's observation can apply to staff officers as well as line officers, or to combat support and combat service support branches in the Army as well as to the combat arms. That debate need not be entertained here. If we accept Lasswell's point that the military deals with violence, we must, in addition, accept this discerning judgment of Sir John W. Hackett:

The essential basis of the military life is the ordered application of force under an unlimited liability. It is the unlimited liability which sets the man who embraces this life apart. He will be (or should be) always a citizen. So long as he serves, he will never be a civilian.[28]

Janowitz puts it succinctly: "The professional soldier always fights."[29] Here, then, is the kernel—the very soul—of the military. They are specialists on violence; they fight under the clause of unlimited liability; they always fight (or, we might concede, are always prepared to fight). Notice that these three elements distinctly parallel Huntington's notions of the professional: expertise (specialists), responsibility (unlimited liability—as the military Code of Conduct requires of the soldier: "I am prepared to give my life in their [my country and our way of life] defense"), and corporateness (the confraternity of the military marked by ritual, uniform, and prescribed behavior).

What imparts force and focus to the preceding is the simple, yet elegant, oath of office. On those occasions when I have been privileged to administer this oath to new officers, I have worn the very uniform in which I served my last day on active duty, creating a bond, however fleeting and tenuous, between the new officer and me. That oath merits quotation:

I do solemnly swear [or affirm] that I will support and defend the Constitution of the United States against all enemies, foreign and domestic; that I will bear true faith and allegiance to the same, and that I take this obligation freely, without any mental reservation or purpose of evasion; and that I will well and faithfully discharge the duties of the office on which I am about to enter. So help me God.

There is the sacred character of an officer's obligation; there is the pious affirmation that distinguishes the military profession.

The venerable West Point motto of "Duty, Honor, Country" is, in fact, the heart of the military ethic. But a word of caution is in order. When Hartle writes that "An officer's loyalty is to the principles and values manifested in the constitution, not to the person of the commander-in-chief,"[30] he (presumably unintentionally) corroborates MacArthur's stand against Truman in 1951. Note the dilemma here: we cannot unreservedly advance the thesis that soldiers must always carry out orders presented them by their chain of command, for we know that there can be illegal and unethical orders. But neither can we unreservedly endorse the notion that an officer's premier loyalty is to his conception of Constitutional imperative, for that way lies anarchy—or an Oliver North-like estimation of governmental powers and of his own broad latitude. *Semper Fidelis* is the worthy motto of the Marine Corps. But always faithful to whom or to what? To the chain of command? No. For illegal orders must be disobeyed. To one's own conception of Constitutional concepts, however Machiavellian or bizarre they may be? No. For ours is a government of laws and not of men (as we are wont to say). And if our soldiers invariably act as self-appointed overseers of the public trust, marching to the beat of a different drummer (one found only in their own minds), we will not have a military profession.

But what appears to be an intractable problem yields to the answer of the military ethic: integrity. The sound moral character of the officer is the sword that cuts the Gordian knot of what I call the "loyalty dilemma." Rarely will there be conflict, after all, between the soldier's loyalty to his commander and that soldier's loyalty to what he perceives as his Constitutional or principled duty. The presumption must always be that the soldier will obey his commander—unless right reason and sound conscience oblige the soldier to refuse. *Conscience*, of course, literally means "with knowledge." Such knowledge, such "right reason and sound conscience," are the product and process of good education, which is the parent of the authentic military ethic. The officer and the patriot are at one. In 1872, Carl Schurz, in a speech to the Senate, properly defined *patriotism*; he could have been speaking, as well, of the loyalty dilemma, or of the American military ethic: "Our country, right or wrong! When right, to be kept right; when wrong, to be put right!" But how do we know when? Look to your own well-formed conscience and peer into the depths of your own soul. There will the answer be found.[31]

IV

If we understand, however intuitively, the military ethic, what of the *American* part? As part of our intellectual and ethical treasury, we are heirs to a tradition of diplomacy that insists that treaties are binding. (Students of diplomacy and international law will recognize the Latin clause *Pacta sunt servanda.*) This is not to suggest that western diplomacy and statecraft are not blemished by multiple episodes of shameful conduct and decision; they are indeed. It is, however, to submit that the exceptions here suggest the viability and the urgency of a general rule of honorable diplomatic conduct. In the early eighteenth century, for example, François de Callieres suggested that "The wise and enlightened negotiator must of course be a good Christian"; and that "a lie always leaves a drop of poison behind, and even the most dazzling diplomatic success gained by dishonesty stands on an insecure foundation."[32] In much the same vein, the English diplomatist Sir Harold Nicolson wrote that the first virtue of the diplomat is truthfulness. His description of the ideal diplomatist captures the western heritage:

> These, then, are the qualities of my ideal diplomatist. Truth accuracy, calm, patience, good temper, modesty, and loyalty. They are also the qualities of an ideal diplomacy.
>
> "But," the reader may object, "you have forgotten intelligence, knowledge, discernment, prudence, hopitality [*sic*], charm, industry, courage and even tact." I have not forgotten them. I have taken them for granted.[33]

These general prescriptions, contrasted with the Byzantine practices of countries in which such advice would be seen as silly or credulous, form the foundation for the western statecraft in which the United States participates. It is beyond the scope of this book to explore this theme at length. Let two brief examples serve the point made here that, generally, Americans expect that Americans will be forthright and honest in their negotiations, in much the sense that President Woodrow Wilson urged. (This does not exclude, by the way, closed-door diplomacy or realism in negotiation.) When President Nixon resigned the presidency in 1974 (facing, entirely properly, the threat of imminent impeachment and probable conviction), people in scores of countries across the globe wanted to know how many people Nixon had killed. When told that, in fact, he hadn't killed anyone but had lied and had covered up a theft, these people were left stunned! This is not to defend Nixon's reprehensible conduct; but consider his crime in the context of Stalin, of Mao Tse-tung, of Hitler, of Tojo, of—the list goes on and on. Consider the manner in which Americans treat their prisoners of war. Compare that with the treatment received by American POWs at the hands of the North Koreans, the North Vietnamese, the Iraqis.

In the earliest days of the war against Iraq, American bombers dropped a greater tonnage of bombs against the Baghdad area than it did against Dresden in 1945—but the raids, clearly, were not designed to perpetrate mass killings of Iraqi civilians. Thus did General Norman Schwarzkopf say of Operation Desert Storm that "We have consistently made a point all along of not going against targets that were not of a military nature."[34] This should be compared with the tactics employed by Saddam Hussein in the early days of that war—bombing innocent civilians in a country (Israel) with which he was not then at war, releasing oil into the Persian Gulf to produce an ecological disaster, employing terrorist tactics whenever and wherever possible. Indeed, at the end of the 1991 war, U.S. forces could have obliterated the Iraqi army fleeing, practically defenselessly, from Kuwait; but President Bush stayed the executioner's hand (as he surely should have). The American approach to warfare, in keeping with the ideals of the "just war," stands in dramatic contrast with that of the vanquished Iraqis.

Much of what can be said about the *American* in the AME sounds either sophomoric or jingoist. Hartle suggests that American core values are freedom, equality, individualism, and democracy.[35] A relevant U.S. Army Field Manual[36] suggests that the "larger moral, spiritual, and social values" upon which our country was founded include "truth, life, liberty, pursuit of happiness, justice and fairness, peace and security, equality of opportunity, and responsibility." The sophisticate may smile a bit cynically at this listing, but the manual's point about these values is clear and keen: these values "provide the basic moral justification for the armed forces . . . [and] the Army's ethic must be consistent with the national will, purpose, and ethic from which it flows." That is exactly correct. All of this will appear terribly qualitative to analysts increasingly given to quantitative analysis. But as Professor Wakin suggests in a quotation he attributes to Milton: "When we can't measure the things that are important, we ascribe importance to the things we can measure."[37]

At least since Plato wrote the *Crito*—and now through William F. Buckley's book *Gratitude*, suggesting that we do indeed owe something to our country—some have realized that much of what we are, we are because of our country. That is true of us as individuals. That is true of our diplomacy. That is true of our military ethic. We cannot define, with perfect precision, the term *American*; nor can we delineate, exactly, the values comprehended by that term. But, somehow, we know those values and what they mean. In fact, the more important something is, the harder it is to define. (I can define, say, *table* easily; I cannot easily define *love* or *justice*.) It is ironic that the more difficult something may be to define exactly, the more important it may be to us! Wisdom and courage and temperance and truthfulness are hard to define; but we know we want those qualities in our friends; and, as de Callieres and Nicolson said, we want them in our diplomats; and we want them, too, in our military officers. They are the classical cardinal virtues, and they help us judge both men and nations. In the United States we are free to employ our own judgment, not required to use that of the government. That is no small thing.

V

"The function of the profession of arms," wrote Sir John Hackett, "is the ordered application of force in the resolution of a social problem."[38] But a great deal of emphasis has been placed in this book on the ideals of liberal education and honor. Some would find those ideals to be wholly inconsistent with the armed forces, which are, after all, agents of death and destruction. As Sam Sarkesian and Thomas Gannon have pointed out, however, "[w]ithout honor and trust the professional soldier is merely a trained killer. With honor and trust, he becomes a man of 'gentlemanly conduct,' professing loyalty to a person or group, devoting himself to 'expertise, responsibility, and corporateness of the profession of arms.'"[39]

John Silber, president of Boston University, explained the idea this way recently:

> I was once asked by a friend to recommend his son for the U.S. Naval Academy. I asked in response, "Does your son want to be a professional killer?" And he said, "Well, no, of course not. He wants to be a peacemaker." I replied, "Well then, tell him to enter a seminary."
>
> A person not prepared to use his skill, knowledge, techniques, and all the weapons at his disposal for the purpose of killing on behalf of the United States of America when ordered to do so has no business in the military. That is the military's ultimate business. That is the profession of the military.[40]

Janowitz's idea, referred to previously, that the professional soldier always fights is the gist of the argument put forward here. The profession of arms is not to be confused with a social cause, with a religious charity, with an educational institution. To cadets and candidates at the academies, in ROTC, in OCS, the message should be given loud and clear: "Your job is to kill the enemy and take ground [or control the seas, or gain air superiority, etc.]! After your four years here I want you to realize all of your training comes down to that bottom line. That is what your job is as a platoon leader in combat." Precisely that message was given recently by Fred Downs to West Pointers. "Afterwards," Downs wrote, "[he] was admonished: 'Here at West Point we do not call it "killing the enemy." We call it "servicing the target." ' "[41] Downs contends that "Killing is the Army's major function during war. Yet it is the least understood, most ignored, and least discussed aspect of a young officer's training . . . ." Downs, an Infantry OCS graduate and rifle platoon leader in Vietnam (who suffered the loss of most of his left arm during that war), is entirely correct. Few recruitment posters, after all, tout killing as the major incentive toward joining the Army, Air Force, Navy, or Marine Corps. If a young man walked into a recruiter's office and announced, "I want to kill people!" one supposes that he would be viewed as a psychopath and denied enlistment on grounds of being unfit for military service. The intriguing thing is that such a person most likely would indeed be unfit for the military!

The vast majority of those in uniform will never see any combat directly, although a fraction may be indirectly involved with combat (through logistical activity and so on). Very few military officers, or enlisted personnel, spend even five percent of their careers in combat. But every man and woman in uniform is concerned, however obliquely, with death and destruction. This hardly means that all soldiers and sailors are necrophiliacs; it does mean that wearing the uniform obliges one—except in the very extraordinary circumstances of a nascent My Lai—to kill; to be prepared to kill; or to assist, in virtually any way possible, those who are directly involved with killing the enemy. (*Kill*, in a military sense, does not necessarily mean "to cause to die," although, of course it often does mean precisely that; but it can also mean "to cause to cease operating" [as in "to kill an engine"], by which is meant compelling the enemy to suspend his hostile operations and thus, in essence, to accept your will.) But the point is that those in the profession of arms must be prepared to kill. They must also be prepared to die.

VI

Even junior military leaders quickly learn that the military ethic incorporates a two-fold responsibility: to accomplish their mission and to provide for the welfare of their men. Any leader, civilian or military, can be said to have a similar duty: to do his job and to look out for his subordinates (or students or clients or patients, and so forth). It is only in the military, however, in which the two duties can conflict absolutely. The best illustration of this leadership dilemma is the example of Commander Lloyd Bucher. As captain of the *Pueblo*, Bucher in 1968 was confronted with a cruel choice: would he surrender his ship to the North Koreans or resist—possibly at the price of the lives of his men?

Bucher surrendered. Surrounded by North Korean warships, already fired upon by them (resulting in the death of one U.S. Navy crewman), with no chance of offering effective resistance, Bucher reasoned that any attempt to resist would result in more needless deaths of his crewmen. Similarly, he believed that any effort to run would be fruitless and that the *Pueblo* would be sunk in short order by the North Koreans. Thus Bucher chose to regard the welfare of his men as more important than his mission, which, at that time, had become the responsibility of preventing the search and seizure of his ship by the North Koreans. Bucher's ship had had an electronic surveillance mission but, threatened with seizure, that mission immediately became the protection of his ship's very sensitive codes and equipment from capture by hostile forces.[42]

Those not on board the *Pueblo* in January 1968 and those who are not naval experts can hardly hope to judge the actions of Commander Bucher. After all, Bucher was the captain, the man on the spot, and the ultimate decision and responsibility were his alone. Yet, *understood as principle*, Bucher's decision to surrender appears to have been the wrong one. Whereas a civilian should have chosen as Bucher did, a military man's first impulse must be to the accomplishment of the mission and second to consideration of the welfare of his men. Were it to be otherwise, any military operation involving a risk would not be undertaken, and in fact military forces could be dispensed with. It is necessary to add the caveat that, in some circumstances, to increase or to continue military operations can be so obviously futile that only a madman would choose such a course. American service personnel are permitted to surrender when they no longer have the means to resist—a phrase that is and must be vague at best. (Could Bucher have fired his deck gun with any effect on the North Koreans? Could his sailors have repelled boarders effectively? How much more time did his crew need to destroy the classified materials and machines on board?) But the principle of *the mission first* is, within the confines of common sense and the call of duty, the very core of the American military ethic. It is also a professional military imperative that most of civilian society cannot be expected to understand or to agree with.

The tension between the two duties of leadership pales by comparison with the tension between the military and civil society. There is no clearer example of this tension than that provided by Joel Jay Finer, an attorney, who became convinced that the Navy's Board of Inquiry, convened to examine the seizure of the *Pueblo*, was a cruel and tormenting inquisition. Finer strenuously objected to an examination of the seizure by a board of naval officers. He thought it outrageous that Bucher's professional colleagues would be charged with conducting the inquiry, and he argued that a group of selected civilians could have investigated with equal competence but with greater compassion.[43] But Finer is wrong precisely because he fails to recognize the distinction between the military and civilian communities. Attempting to evaluate or judge military discipline from a civilian point of view is a senseless exercise: civilians may suffer hardships and they may even risk their lives but, *ordinarily*, they will do so only when they expect immediate personal gain from their actions. The circumstances in which military personnel often find themselves, however, rarely permit actions calculated to achieve mere personal advantage. And military proficiency, discipline, morale, and esprit—factors always best judged by military professionals themselves—are the means to the end of mission accomplishment. No civilian without military experience or indoctrination can be expected to understand the military *Weltanschauung*, and no civilian can be (or should be) expected to judge alleged departures from it. The rationale for the Board of Inquiry was beyond Finer's ken. And the thought of a court martial of Commander Bucher (as the board recommended) was virtually unthinkable in a liberal society that was, and not without reason, persuaded not only of Bucher's innocence but of his heroism.

One can appear cavalier in saying that military personnel must be prepared to kill—and that they must also be prepared to die. But if we focus on the AME, it teaches us that lesson, however difficult it is to accept. General Ridgway seemed to hit the right note: "It has always seemed to me that a commander has as deep a duty to the men with whose lives he is temporarily entrusted as they have to him—and part of that duty is to see to it that those lives are not needlessly expended."[44] It can also appear glib to say that the military is in, but not of, its parent society; it is a separate confraternity. The democratic society our military exists to protect ceaselessly attempts to proselytize—to convert its military to its democratic philosophy. But whereas the democratic polity is characterized by an emphasis on the individual, by an egalitarian impulse, and by a deliberate or judicial decision-making process, the military model is characterized by an emphasis on the interest of the organization, by a well-defined hierarchical structure, and by a command or authoritarian decision-making process. Procrustean efforts to meld the two models run the real risk of destroying both! As S. L. A. Marshall has said:

> [T]o the extent that military men lose their faith in [discipline's] virtue and become amenable to ill-considered reforms simply to appease the public, they relinquish the power to protect and nurture [their society] . . . . The relation of the military establishment to American democracy is that of a shield covering the body. But no wit of man can make it a wholly "democratic" institution in its processes without vitiating its strength, since it progresses through the exercise of unquestioned authority at various levels.[45]

As Janowitz observed, "To deny or destroy the difference between the military and the civilian cannot produce genuine similarity, but runs the risk of creating new forms of tension and unanticipated militarism."[46] Thus to talk about killing and dying, however understandably unpleasant it may be, is to talk about the military's *raison d'être*. Euphemisms may have their place. But the military's job remains: it is to find, fix, fight, and finish the enemy. Civilians do not do those things. If our armed forces do their military jobs right, civilians will never have to do those things.

VII

In writing the third of the *Federalist Papers*, John Jay, destined to become the first chief Justice of the fledgling U.S. Supreme Court, wrote that "Among the many objects to which a wise and free people find it necessary to direct their attention, that of providing for their *safety* seems to be the first." National safety and security must be the principal focus of the professional throughout his career. When he puts his career ahead of his country, he has failed. The AME insists that the officer's integrity be the cornerstone of his career. An officer who willingly sacrifices principle on the altar of expediency has betrayed the core of the profession he supposedly serves.

The U.S. Army, in its FM 100-1, maintains that "the professional military ethic articulates our values . . . . The ethic sets the moral context for the Army in its service to the nation and inspires the sense of purpose necessary to preserve the nation." Loyalty, duty, selfless service, and integrity are regarded by that manual as "essential values that guide the way we live our lives and perform our duties." The same manual suggests that commitment, competence, candor, and courage are the four "individual values" that "can strengthen the professional Army ethic." It is not clear why that manual segregates those eight values into two camps, but to pursue this one question is a mere quibble. What truly matters is the Army's definition of *courage*. After providing a definition of physical courage—"the ability to overcome fear and carry on with the mission"—the manual contends that "Moral courage, the courage of one's convictions, is equally important." And so it is.

Without moral courage, the professional soldier endangers the very society he is engaged to protect. Let no one think that the obsequious, robot-like acceptance of orders, without question, is the mark of the professional. General Ridgway once pointed out that in a fighting Army composed as it is principally of young, fairly athletic men, physical courage will not prove to be in short supply. But moral courage may well be wanting. Richard Gabriel has often been too glib with his solutions to certain military problems,[47] but he is exactly on target in this asseveration: "The nature of command and military service is a moral charge that places each soldier at the center of unavoidable ethical responsibility."[48] Consider that obligation in light of the comment made by former Army Chief of Staff General E. C. Meyer:

> The obligation of service and commitment inherent in the military ethic imposes burdens not customary in the larger society where obligations are normally contractual in nature and limited in degree of personal sacrifice expected. For the soldier, the obligation is complete: to death if necessary.[49]

That is Hackett's "clause of unlimited liability." If the soldier's liability can be total, his ethical sensitivity should be equally acute.

John Silber has suggested that the soldier "should not fear losing his life, but his honor. He may not be able to preserve his life, but he can always preserve his honor."[50]

We come then full circle. We return to the idea that officers—indeed, all service personnel—must tell the truth. The sergeant who leads a patrol and reports an area cleared of the enemy when his patrol was never in that area endangers fellow soldiers of his platoon or company or battalion. A general who lies may well endanger the nation. Silber again:

> A four-star general headed for the chairmanship of the Joint Chiefs of Staff may be asked by the commander-in-chief for his best advice on a professional issue with serious political ramifications. The general may suspect that the commander-in-chief wants advice that is contrary to the general's best professional judgment, and that to state his true opinion may lead to premature retirement. The history of Vietnam records examples of presidential importuning of just this sort. But the general should not fear the termination of his career; rather, he should fear the dishonor of denying the commander-in-chief his best professional judgment. Morally, such a failure would be indistinguishable from abandoning one's post under fire. *The soldier should be prepared not only to die for his country, but to be fired for it.* [emphasis added.][51]

VIII

The AME is based upon personal and *operational* integrity. The two are entirely symbiotic. A commander who is a moral weakling is likely to blunder into war crimes. Operations planned without sustained attention to the Rules of Engagement (ROE), normally worked out to combine tactical security with the exigencies of the Laws of War, are likely to work incalculable hardships on commanders in that area of operations. This is not the forum for lengthy discussions of ROE or the idea of the just war. Colonel Hartle points out that "The most severe critics of just-war theory are radical pacifists who reject all use of force in response to violence."[52] Only the fool can believe in total war. Only the fool can believe in total peace. There is such a thing as remorseless violence. But there is also such a thing as *remorseless peace*. We too often forget that the man (or nation) prepared to stand idly by while another man (or another nation) is ravished is just as likely to stand by when our turn comes.

The American military ethic is in keeping with the goals and suppositions of classical just war theory. The just war requirements are divided into the *Jus ad Bellum* (concerning going to war) and the *Jus in Bello* (concerning waging the war). To go to war, the decision must be made by competent authority (not by some wild-eyed colonel); the cause of the war must be just and prospects for settlement by other means exhausted (the war must be a virtual last resort); and the war must be waged with right intention (its purpose is to be a just and lasting peace). To conduct the war, the operations should be conducted in accordance with the principle of proportion (meaning consistency between political-military ends and the military means); they should be conducted in accordance with the principle of discrimination (which prohibits intentional and direct attacks on noncombatants and non-military targets); genocide, of course, is prohibited; and the published laws of war (such as treaties) must be observed.[53] As *Time* put it: "If there is such a thing as a just war, President Bush launched one against Saddam [in January 1991]."[54]

The manner in which we go to war and wage it—remarkably open for all the world to see—is a striking comment about the American military ethic. Our democratic decision in 1991 to employ force; our exhausting practically every diplomatic and economic prospect for peace before resorting to violence; our intention to help build a lasting peace in the Middle East (a Sisyphean labor if ever there were one!); our effort to bomb legitimate military targets and to spare civilian targets insofar as humanly possible; our observation of strict ROE and precise treatment of POWs; our reluctance to launch all-out ground operations with our troops until enemy targets had been softened up; our willingness to end the war without wholly destroying Saddam's defeated armed forces—all these things speak volumes about the military ethic professed and practiced by American commanders.

There will ever be need of study and review of the tenets of just war.[55] There will ever be need of study and review of the internal operations of the U.S. military. (Northwestern sociologist Charles Moskos observed recently that "There is probably a better racial climate in the military than on a college campus."[56]) Twenty years ago, the military appeared to be in a kind of death-agony. A resurrection of sorts appears to have taken place. And the reasons for it are clear: in addition to the military triumph over the Iraqis, the traditional values of the professional are again ascendant. The AME is well because it has been examined and evaluated, discussed and debated. The critics have criticized—and, in turn, have been criticized. That is our national way. In the end, the American military ethic will remain truly *professional* so long as our armed forces personnel and, indeed, all Americans think of their military in terms of a World War II idea (and expression) too long neglected. In World War II, especially, when one joined (or was drafted into) the Army, Navy, or Marine Corps, he thought of himself as being—and was said to be—*in the service*. How perfectly concise! In the service. After all is said and done, that simple prepositional phrase captures it all. Those who wear the uniform are *in the service*.

The third stanza of *The Star-Spangled Banner*, too rarely sung and almost unknown, provides this perspective on our country and its military forces:

> O thus be it ever when free men shall stand
> Between their loved homes and the war's desolation;
> Blest with victory and peace, may the heaven-rescued land
> Praise the Power that hath made and preserved us a nation.
>
> Then conquer we must, when our cause it is just,
> And this be our motto, "In God is our trust."
> And the Star-spangled Banner in triumph shall wave
> O'er the land of the free and the home of the brave.

The cynic will consider that to be mawkish. But, in fact, it is a concise and cogent invocation—and a majestic reminder that the American military ethic, no less than the *Star-Spangled Banner*, is the national anthem.

NOTES

1 Sir John W. Hackett, "Profession of Arms," *Military Review* 43 (October 1963): 44.

2. Sidney Hook, *Marxism and Beyond* (Totowa, NJ: Rowman and Littlefield, 1983), p. 207.

3. John W. Hackett, in a 1970 lecture at the Air Force Academy, quoted in John F. Reichart and Steven R. Sturm, *American Defense Policy*, 5th ed. (Baltimore, MD: The Johns Hopkins University Press, 1982), p. 737.

4. Max De Pree, *Leadership is an Art* (New York: Doubleday, 1989), p. 17. This quotation is in italics in the original text.

5. James Bond Stockdale, "Educating Leaders," in *Military Leadership: In Pursuit of Excellence*, ed. by Robert L. Taylor and William E. Rosenbach (Boulder, CO: Westview Press, 1984), p. 67. Stockdale's reference to Monod is doubtlessly to Gabriel Monod (1844–1912), an historian who helped introduce German historical methodology to France. Considered to be among the most scholarly and

stimulating teachers of history in his day, Monod also greatly contributed to the seminar system.

6. Robert G. Gard, "The Military and American Society," *Foreign Affairs*, 49 (July 1971): 710.

7. Brogan is quoted in Anthony L. Wermuth, "A Critique of Gabriel and Savage," *Armed Forces and Society* 3 (May 1977): 482.

8. Douglas MacArthur, "Duty, Honor, Country," in *A Soldier Speaks*, ed. by Vorin E. Whan, Jr. (New York: Praeger, 1965), p. 357.

9. Samuel Eliot Morison, *The Oxford History of the American People* (New York: Oxford University Press, 1965), p. 1072.

10. Samuel Huntington, *The Soldier and the State* (Cambridge, MA: Belknap Press, 1957), p. 351.

11. As Paul Johnson points out in *Modern Times*, "The Americans could not understand," General Montgomery wrote sadly, "that it was of little avail to win the war strategically if we lost it politically." Johnson contends that Eisenhower so wanted Red Army cooperation against Japan that eastern Europe and most of the Balkans were lost to the Soviets. (New York: Harper and Row, 1983), p. 436.

12. Morison (above, n. 9), p. 1065.

13. Jerome Slater, in Reichart and Sturm (above, n. 3), p. 750.

14. See John Lovell's astute criticism of the Slater article in ibid., pp. 756–760.

15. Sun Tzu, *The Art of War*, ed. by James Clavell (New York: Delacorte Press, 1983), p. 9.

16. John H. Garrison, in Reichart and Sturm (above, n. 3), 763.

17. Garrison, ibid., p. 765.

18. Robert F. Kennedy, *Thirteen Days* (New York: W. W. Norton, 1969), p. 119.

19. Lovell in Reichart and Sturm (above, n. 3), p. 759.

20. Lloyd J. Matthews, "Resignation in Protest," *Army*, January 1990: 18.

21. Huntington (above, n. 10), pp. 8–10. See also Reginald J. Brown, "The Meaning of Professionalism," *American Behavioral Scientist* 19 (May/June 1976): 511–512.

22. Morris Janowitz, *The Professional Soldier* (Glencoe, IL: The Free Press, 1960), p. 215.

23. Anthony Hartle, *Moral Issues in Military Decision Making* (Lawrence, KS: University Press of Kansas, 1989), p. 27.

24. Quoted in MacArthur (above, n. 8), p. 69.

25. Alfred Thayer Mahan, "Naval Education," *U.S. Naval Institute Proceedings*, 5 (1897): 347.

26. Hartle (above, n. 23), p. 8.

27. Harold Lasswell, *The Analysis of Political Behavior* (New York: Oxford University Press, 1949), p. 146. This is a reprint of his famed 1941 essay, "The Garrison State and Specialists on Violence," from the January 1941 issue of *The American Journal of Sociology*.

28. Sir John W. Hackett, "Profession of Arms, II," *Military Review* 43 (November 1963): 57. I cite this journal, for it may be more readily available than Hackett's superb book, from which *Military Review* printed their two articles thirty years ago. See Hackett's excellent book *The Profession of Arms* (London: Times Publishing, 1963); Macmillan reprinted it in 1983.

29. Janowitz (above, n. 22), p. 215.

30. Hartle (above, n. 23), p. 44.

31. See, for example, John F. Kennedy, *Profiles in Courage* (New York: Harper, 1956) and Walter Lippmann, *Essays in the Public Philosophy* (Boston: Little, Brown, 1955).

32. François de Callieres, *On the Manner of Negotiating With Princes*, trans. by A. F. Whyte (Notre Dame, IN: University of Notre Dame Press, 1963), pp. 31, 126.

33. Sir Harold Nicolson, *Diplomacy*, 3rd ed. (New York: Oxford University Press, 1964), pp. 58, 67.

34. *U.S. News & World Report*, 28 January 1991: 14, 31–32. See also "After the Storm," *Newsweek*, 11 March 1991: 26–29.

35. Hartle (above, n. 23), p. 85. To explore this concept more rigorously, see Daniel J. Boorstin, *The Americans: The Democratic Experience* (New York: Random House, 1973) as a beginning point.

36. FM 100-1, Headquarters, Department of the Army (Washington, DC: Government Printing Office, 1986), p. 21.

37. Malham M. Wakin, "The Ethics of Leadership, II," in *War, Morality, and the Military Profession*, ed. by M. M. Wakin, 2nd ed., revised (Boulder, CO: Westview Press, 1986), p. 210.

38. Sir John W. Hackett, *The Profession of Arms* (London: Times Publishing, 1963), p. 3.

39. Sam C. Sarkesian and Thomas M. Gannon, "Introduction: Professionalism—Problems and Challenges," *American Behavioral Scientist* 19 (May/June 1976): 504. They cite Donald F. Bletz, *The Role of the Military Professional in U.S. Foreign Policy* (New York: Praeger, 1972).

40. John Silber, *Straight Shooting* (New York: Harper and Row, 1989), p. 255.

41. Fred Downs, "It's Time West Point Stopped Sugarcoating Combat," *The Washington Post National Weekly Edition*, 31 August 1987: 23. See also Downs's book *Killing Zone* (New York: W. W. Norton, 1978).

42. For details, consult Lloyd Bucher, *Bucher: My Story* (Garden City, NY: Doubleday, 1970); his executive officer was Edward Murphy, whose view of events is somewhat different from his skipper's: *Second in Command* (New York: Holt, Rinehart, Winston, 1971); and Daniel V. Gallery, *The Pueblo Incident* (Garden City, NY: Doubleday, 1970).

43. Joel Jay Finer, "The Second Ordeal of the *Pueblo* Crew," *American Bar Association Journal* 55 (November 1969): 1029–1033.

44. Matthew B. Ridgway, *The Korean War* (Garden City, NY: Doubleday, 1967), p. 63.

45. [S.L.A. Marshall], *The Armed Forces Officer*, DOD Pam. 1-20 (Washington, DC: Government Printing Office, 1960), p. 160.

46. Janowitz (above, n. 22), p. 440.

47. See especially Matthews (above, n. 20), pp. 15, 21.

48. Richard A. Gabriel, *To Serve With Honor* (Westport, CT: Greenwood Press, 1982), p. 140.

49. Quoted in M. M. Wakin (above, n. 37), p. 211.

50. Silber (above, n. 40), p. 258.

51. Ibid.

52. Hartle (above, n. 23), p. 58.

53. Discussion based upon William V. O'Brien, *The Washington Papers. U.S. Military Intervention: Law and Morality*. No. 68 (Beverly Hills, CA: Sage, 1979), p. 38.

54. *Time*, 28 January 1991: 16.

55. See Douglas Lackey, *The Ethics of War and Peace* (Englewood Cliffs, NJ: Prentice-Hall, 1989). Lackey's book contains a number of excellent suggested readings and references. Also see Joseph Martino, *A Fighting Chance: The Moral Use of Nuclear Weapons* (San Francisco, CA: Ignatius Press, 1988). Some other references: Michael Walzer, *Just and Unjust Wars* (New York: Basic Books, 1977)—now a standard. George Weigel, *Tranquillitas Ordinis* (New York: Oxford University Press, 1987) is a superbly written volume. An excellent general volume is James E. Dougherty and Robert Pfaltzgraff, *Contending Theories of International Relations*, 3rd ed. (New York: Harper and Row, 1990).

56. *The New York Times*, 25 January 1991: A8. Blacks now make up 30% of all U. S. Army troops. Ibid., p. 1.

RECOMMENDED READINGS

The principal sources used in the writing of this book are acknowledged in the endnotes of each chapter. Those specific references, cited in conjunction with each chapter, are the ones most highly recommended to anyone who wishes to pursue to greater lengths any of the themes mentioned in this volume. The following selections are of a general, background nature and may also be of use or interest.

According to Colonel Harry Summers (U.S. Army, Retired), the Army has been highly trusted by Americans throughout our history. Only recently has that sentiment changed. Perhaps, he suggests, that sentiment has once again begun to change—now to the good. Certainly the successful prosecution of the Persian Gulf War will—as it should—enhance the military's image and standing. For an analysis, see J. Francis Gladstone, "The Military Mind," *The Boston Globe Magazine*, 8 February 1987, p. 14 and following. The preface and introduction to this book attempt to place this volume into the so-called "reform literature" of the past quarter-century. The critics, although sometimes tendentious (and occasionally even vitriolic), were also, let it be said, frequently correct. The danger in any reformation movement, however, is that it can go too far too fast. Not all was as decadent as some critics alleged. See, for example, Daniel P. Bolger, *Americans at War, 1975–1986: An Era of Violent Peace* (Novato, CA: Presidio Press, 1988), especially pp. 443–452 for an interesting rejoinder to some well-known critics. One Vietnam veteran, rather a precursor of sorts to Vann and Hackworth, was the highly decorated Anthony R. Herbert. See his book (written with James T. Wooten, *Soldier* (New York: Dell, 1973) to understand some of the roots of the reform movement, good and bad. A Vietnam-era veteran who criticized the Army was Edward L. King in *The Death of the Army* (New York: Saturday Review Press,

1972). An extraordinary novel of the times is Josiah Bunting, *The Lionheads* (New York: Popular Library, 1972), which tells the story of careerism and reveals how it destroys professionalism. Just as some military men were beginning to suggest changes in the military [e.g., Zeb B. Bradford and Frederic J. Brown, *The United States Army in Transition* (Beverly Hills, CA: Sage, 1973)], civilian journalists and scholars were beginning their own analyses [e.g., Ward Just, *Military Men* (New York: Knopf, 1970)]. Soldier-scholars such as Harry Summers [*On Strategy* (New York: Dell, 1982)] and Andrew Krepinevich [*The Army and Vietnam* (Baltimore, MD: The Johns Hopkins Press, 1986)] were responding to the need for research and reform in a disciplined, properly conservative manner. That measurable success has been achieved seems indicated by Bob Woodward's informative, if rather breezy book *The Commanders* (New York: Simon & Schuster, 1991).

Strategy and war have never been popular studies in the academy, and that is extremely unfortunate. The unpopularity of military history at many universities is regrettable—and reminiscent of the twisted logic of refusing to permit ROTC to be offered on prestigious campuses. Are not the graduates of such institutions precisely the kind of people (in many respects) we want in our military? In any event, the literature of strategy and war is important and remarkably rich. As an example, consult such works as Bernard Brodie, *War and Politics* (New York: Macmillan, 1973); Michael Howard, *War in European History* (New York: Oxford University Press, 1976); Basil Liddell Hart, *Strategy*, 2nd rev. ed. (New York: Praeger, 1967); Peter Paret, ed., *Makers of Modern Strategy* (Princeton, NJ: Princeton University Press, 1986); George E. Thibault, *The Art and Practice of Military Strategy* (Washington, DC: National Defense University, 1984); and Harry R. Borowski, *The Harmon Memorial Lectures in Military History, 1959–1987*, Office of Air Force History (Washington, DC: Government Printing Office, 1988). Three good general studies are Geoffrey Blainey, *The Causes of War* (New York: The Free Press, 1973); John G. Stoessinger, *Why Nations Go to War*, 5th ed. (New York: St. Martin's Press, 1990); and the late Paul Seabury and Angelo Codevilla, *War: Ends and Means* (New York: Basic Books, 1989). These volumes contain, of course, invaluable references and notes for those interested in further reading and research.

Scholars at work on military issues often forget our responsibility to teach. In listing page after page of references we want to read, we forget that, to undergraduates just beginning in these fields, such lists can appear daunting, to say the least. Textbooks are not popular among scholars, for they reiterate the things most of us knew a long time ago. The case is very different, however, with undergraduates. Classic and beautifully written works such as Inis Claude's *Power and International Relations* (New York: Random House, 1962) and Kenneth Waltz's *Man, the State, and War* (New York: Columbia, 1959) can still be tough sledding for the tyro. A well-crafted text, such as John Spanier, *Games Nations Play,* Seventh ed. (Washington, DC: Congressional Quarterly Press, 1990) or Theodore A. Couloumbis and James H. Wolfe, *Introduction to International Relations,* Fourth ed. (Englewood Cliffs, NJ: Prentice-Hall, 1990) or Michael Roskin and Nicholas Berry, *IR: An Introduction to International Relations* (Englewood Cliffs, NJ: Prentice-Hall, 1990), can be critical to attracting the long-term interest of the neophyte.

One of the best things teachers can do is to get students to shelf read in the library.

College courses are not green stamps pasted into stamp books that one then turns in for a degree. At least that should not be the case! Courses introduce materials; they do not *conclude* such materials. In smaller libraries, the Dewey Decimal system is used. Students should be encouraged to browse. In international relations, for example, students should look through their library's collection under numbers in the 327 range. International conflict is at 327.17. Political ethics is at 172. Professional ethics at 174. International law is at 341. American history at 973.

In the area of leadership and management (again—aside from the references in the endnotes), one might consult such sources as John G. Meyer, Jr., *Company Command: The Bottom Line* (Washington, DC: National Defense University Press, 1990). ROTC courses, with rare exception, are bland, generic, and dull. If ROTC instructors and regular faculty get together, under the guidance of a creative PMS, some very imaginative courses can result. Students and cadets can profit a great deal from reading good books about leadership and management, always asking the question, "What does this book teach *me*?" Kenneth Blanchard and Normal Vincent Peale, *The Power of Ethical Management* (New York: Morrow, 1988) is a very concise, readable place to start. Useful background books would include C. Robert Kemble, *The Image of the Army Officer in America* (Westport, CT: Greenwood Press, 1973), containing an excellent bibliography; and Peter Karsten, *Soldiers and Society: The Effects of Military Service and War on American Life* (Westport, CT: Greenwood Press, 1978), with useful background materials about recruitment practices and basic training.

I have in this book tried to stay away from certain sociological issues beyond the scope of my present effort. Brian Mitchell's book, *Weak Link: The Feminization of the American Military* (Washington, DC: Regnery Gateway, 1989) is somewhat less circumspect: "The American public is being lulled into the mistaken belief that women can, indeed, perform as well as men in all military jobs" (p. 225). Compare Charles Moskos, "Army Women," *The Atlantic*, August 1990: 70–78. Because 11% of the Army's personnel is female, this issue will become increasingly important. See, for example, the discussion in *Time*, 12 August 1991, p. 31.

Education is the key to leadership excellence—that is a principal point of this book! Yet modern education, if we are to believe recent press notices, is an abject failure. Too many colleges countenance too many "candy courses." There are even plenty of "candy majors." See Russell Kirk, *Decadence and Renewal in the Higher Learning* (South Bend, IN: Gateway, 1978) for a scathing critique of higher education. Allan Bloom's book, *The Closing of the American Mind* (New York: Simon and Schuster, 1987) said much the same as Kirk had done a decade before. E.D. Hirsch's book, *Cultural Literacy* (Boston, MA: Houghton Mifflin, 1987) is well worth reading. When I am (if only peripherally) involved in hiring (or in promotion/tenure) decisions, I look for competence, organization, fairness, and enthusiasm in the candidates whom I am assessing. A key consideration (other than integrity and a sense of humor) is the person's commitment to reading. If he/she reads very widely and has some intelligent note-taking or retrieval system, chances are that that person will continue to grow. To teach, one must love to learn; to lead, one must read. Every person worth his salt is an autodidact.

Help for autodidacts is available from Mortimer Adler, *A Guidebook to Learning* (New York: Macmillan, 1986) and his book *The Paideia Proposal* (New York:

Macmillan, 1982) is also excellent. The best book on teaching I know is Gilbert Highet, *The Art of Teaching* (New York: Knopf, 1950). Among the books recommended by Admiral Stockdale when he was teaching at the Naval War College was Homer W. Smith, *Kamongo* (New York: The Viking Press, 1932). It concerns evolution, progress, moral development, purpose, and so on. Why is that on a reading list at the Naval War College? If I haven't already answered that question to the reader's satisfaction, I guess I should have another try at discussing education and the military.

I will violate my own guideline and recommend again here, even though it appears already a couple of times in the endnotes, Roger H. Nye, *The Challenge of Command: Reading for Military Excellence* (Wayne, NJ: Avery Publishing Group, 1986). Glenn Tinder's book *Political Thinking: The Perennial Questions*, multiple editions (Boston, MA: Little, Brown) is an excellent companion volume. Hans Morgenthau wrote a number of superb books, but *Scientific Man vs. Power Politics* (Chicago, IL: University of Chicago Press, 1946) is my favorite. Eric Voegelin's classic *The New Science and Politics* (Chicago, IL: University of Chicago Press, 1952) is a good place to head after you have the basics of political theory in mind. "What has all this got to do with war?" one could ask. Is the idea of just war important? Why do armies fight? What is worth living for or killing for? There is not only a connection between warfare and political theory—the connection is truly vital.

Students of the military *must* read history and biography. Once again, students should browse through the biography section in their public or college libraries. Eric Larrabee, *Commander-in-Chief* (New York: Harper and Row, 1987) has some interesting biographies of FDR's major commanders, although I did not like his views about MacArthur. For military leaders, reading in the standard professional journals is *de rigueur*. I will not list them here, but I add to their number *Armed Forces and Society*, which military professionals should subscribe to (and read!). Vol. 17 (Winter 1991) of that journal, for example, explains (in an article by T. J. Gough) the "Origins of the Army Industrial College" (259–275). This journal also contains a number of useful book reviews in each issue. Reading book reviews, by the way, is important, for reviews not only commend and criticize books but announce their existence to you!

At the Air War College, I teach a course I call "Command and Conscience." One of my classmembers said he was taking it because "Ethics is an 'in' thing." I said, "Colonel, when ethics isn't the 'in' thing, sooner or later everyone will know it." Reading about ethics doesn't make one ethical; but it can increase our awareness and sensitivity. There are a number of references one can consult with high profit. Here are a few: Guenter Lewy's book *America in Vietnam* (New York: Oxford University Press, 1978) caused a storm of controversy when it appeared because a number of people, mostly to the left of the spectrum, did not like Lewy's argument that the Rules of Combat Engagement in Vietnam were moral and highly restrained. Although Lewy was no champion of the military—he was a scholar following the evidence as he read it—he did not denounce the U.S. military as "butchers," which a number of his adversaries "knew" had to be true. His book stands vindicated in my view. The controversy is a fascinating one. Peter Karsten, *Law, Soldiers, and Combat* (Westport, CT: Greenwood Press, 1978) is an interesting volume with useful references. Lloyd Matthews and Dale Brown edited *The*

Parameters of Military Ethics (Washington, DC: Pergamon-Brassey's, 1989). *Military Ethics* is the title of a 1987 National Defense University (Washington, DC) book; in 1988 the same press produced *Moral Obligation and the Military: Collected Essays*. Barrie Paskins and Michael Dockrill, *The Ethics of War* (Minneapolis, MN: University of Minnesota Press, 1979) is very well done. And Peter L. Stromberg, Malham Wakin, and Daniel Callahan, *The Teaching of Ethics in the Military* (Hastings-on-Hudson, NY: The Hastings Center, 1982) has the best annotated bibliography in this field I have seen (pp. 77–85). An extraordinary book is that of Elmo Zumwalt, *My Father, My Son* (New York: Macmillan, 1986). If anyone wants to understand *duty* and *loyalty*, he should read this book. An excellent journal in this general field, by the way, is *Ethics & International Affairs*.

Civil-military relations are our "bright, shining truth," I have argued. To pursue this, see Charles W. Ackley, *The Modern Military in American Society* (Philadelphia, PA: The Westminster Press, 1972); Charles L. Cochran, *Civil-Military Relations: Changing Concepts in the Seventies* (New York: The Free Press, 1974); James Clotfelter, *The Military in American Politics* (New York: Harper and Row, 1973); Henry E. Eccles, *Military Concepts and Philosophy* (New Brunswick, NJ: Rutgers University Press, 1965); and Amos Perlmutter, *The Military and Politics in Modern Times* (New York: Yale University Press, 1977). Two classics are those of J. Glenn Gray, *The Warriors: Reflections on Men in Battle* (New York: Harper Torchbook, 1970) and James Turner Johnson, *Just War Tradition and the Restraint of War* (Princeton, NJ: Princeton University Press, 1981). I have just finished reading John Finnis, Joseph Boyle, and Germain Grisez, *Nuclear Deterrence, Morality, and Realism* (New York: Oxford University Press, 1987); it is an excellent contribution to a number of fields.

INDEX

Adams, Samuel A., 220–221
Advanced Individual Training (AIT), 28–35
Air Command and Staff College, 180
Air Force Institute of Technology (AFIT), 181
Air War College, 177, 180–183, 192
Airborne School, 59–63
Alcohol abuse, 100–103, 108–111
Armed Forces Staff College, 180
Army Command and General Staff College, 175, 180
Army War College, 179–180, 185

Barracks thief, 112–113
Basic Combat Training (BCT), 15, 18–27
Basic Rifle Marksmanship (BRM), 21–22
Bayonet training, 23
Berlin wall, 12
Bible, quotations from, 65, 123, 175, 239
"Biloxi Blues" (Neil Simon play), 27
Bishop, Joseph W., Jr., 208–209
Bradley, Omar, 169–170
Bucher, Lloyd, 246–247
Bundy, McGeorge, 161
Bush, George, 216, 251

Calley, William, 2, 46–47, 215
Capstone course, 181, 196
Careerism, 82–86, 142–144
Carr, E. H., 1
Casey, William, 222
Chemical training (CBR/NBC), 23–24
Churchill, Winston, 1
Cicero, 168–169
The Citadel, 155–158, 177, 194
Compromise, definition of, 98
Convoy, *see* nuclear weapons convoy
Coolidge, Calvin, 201
Cuban missile crisis, 236

de Tocqueville, Alexis, 224–225
Death of fellow officer, 81–82
Debts ("just"), 88–89
Decatur, Stephen, 13

"Democratic soldier," idea of, 201–225
Demosthenes, 153
DePree, Max, 65, 103, 120, 232
DEROS (date eligible for return from overseas), 108, 149–150
"Desert Storm," 204–206, 242–243, 251
Dickey, John Sloane, 153
Disciplinary problems, 129–132
Dixon, Norman, 187–189
Documents, secret, 105–108
Dog handlers, 108–111
Downs, Fred, 244–245
Drill sergeant, 19, 67
Drugs, 104, 126–129; atropine abuse, 140
Dugan, Michael J., 223
Dying and the military, 245–248

Education, definition of, 159–160
Eisenhower, Dwight D., 205–206
Emerson, Ralph Waldo, 66
The enlightenment, 201
Enlisted Efficiency Report (EER), 147–148
Enlisting in the Army, 10
Ethic, definition of, 238

The Federalist, 225, 248
"Feres Doctrine," 213
Fonda, Jane, 155

Gabriel, Richard A., 3–4, 186–187, 249
Galvin, John R., 195
Gannon, Thomas, 244
Goldwater-Nichols (Department of Defense Reorganization Act), 178
"The Graduate" (movie), 14
Grenade training, 22

Hackett, John W., 231–232, 239–240, 244, 249
Hackworth, David, 5, 205
"Hanoi Hilton," 154–155
Harlan, John M., 213
Hartle, Anthony, 238–240
Homosexual issue, 105–107
Hook, Sidney, 231
Humorous episodes, 90–94, 116–119,
Huntington, Samuel, 188, 202, 206
Hussein, Saddam, 243

Industrial College of the Armed Forces (ICAF), 180
Integrity in reporting, 220
Iran-Contra affair, 221–223

Janowitz, Morris, 238, 240
Jefferson, Thomas, 159
Jones, Bruce, 221
Just war theory, 250–251

Katyn Forest, 12–13
Kennedy, Robert F., 236
Killing and the military, 245–248
Kipling, Rudyard, 168
Kirk, Russell, 166, 233
Kitchen police (KP), 17, 48
Knight, Charles, 97
Knights Templars, 37
Koppel, Ted, 165

Lasswell, Harold, 239
Lavelle, John D., 217–220
LAW (light anti-tank weapon), 29–30
Leadership, essence of, 53–55, 68, 73, 79–81, 116; see also "Negative leadership"
Leadership reaction course, 51–53
Lee, Robert E., 58, 158
Leo, John, 162
Lewis, C. S., 162
Liddell Hart, Basil H., 183–184
Lippmann, Walter, 164
Lombardi, Vince, 97
Loyalty dilemma, 240–241

MacArthur, Douglas, 9, 215, 238
Mahan, Alfred Thayer, 183, 239
Marine Corps Command and Staff College, 180
Marshall, George C., 123, 158, 184
Marshall, S. L. A., 7, 124, 248
Matthews, Lloyd, 237
Meyer, Edward C., 249

"Mickey Mouse," 74, 127, 130
Military failure, causes of, 190
Military industrial complex, 205–206
Milton, John, 168, 195
Mission first, idea of, 247
Modernist model, 234–238
Morgenthau, Hans J., 4, 164–165, 169, 203
Morrill, Justin S., 156
Moskos, Charles, 251
Mylander, Maureen, 184–185, 191

National War College, 179–180, 193
Naval Command and Staff, College of, 180
Naval Post Graduate School, 181
Naval Warfare, College of, 179–180
Negative leadership, 29–33, 43–44, 49, 79, 80, 83–86, 132
Newman, John Henry Cardinal, 160–162, 167
Nicolson, Harold, 242
Nixon, Richard M., 13, 201
North, Oliver L., 221–223
North Georgia College (NGC), 157
Norwich University, 156–157, 161
Notre Dame, University of, 154–155
Nuclear weapons convoy, 134–142
Nye Committee, 205

Oath of Commission, 240
O'Callahan v. Parker, 211, 213
Officer Candidate School (OCS), 30–31, 37–58; graduation from, 57–58; introduction to, 40–42
Officer Efficiency Report (OER), lying in, 84–85
Officer immaturity in OCS, 44–46
Open door policy, 75, 127

Partridge, Alden, 156, 170, 207
Patton, George S., 37
Pay officer, 117–119
Penn, William, 37
Petain, Henri Philippe, 46
Physical abuse in OCS, 43–44
Physical limits, 25
Physical training (PT), 27, 72
Plato, 186, 242
Platoon leader (policies and techniques), 73–79, 108–116, 124–125
Platoon Sergeant, 71, 73, 77–78, 98–103
Politics, military involvement in, 233–241
Powell, Colin, 204
Pride, definition of, 98
Profession, definition of, 237
Professional military education (PME), definition of, 176–177; purpose, 182–183, 194, 196
Professional soldier, 86
Psychiatric ward, 117–119
Public speaking by officer, 119–120
U.S.S. *Pueblo*, 246–248
Pugil sticks, 23
Punishment, mass, 18–19
"Purple-suiters," 178

Racism, 71–72, 125–126
Radford, Charles, 216
Ranger problem in OCS, 56–57
Rape, 133–134
Readiness Condition Report (REDCON), 142–143
Reagan, Ronald, 221–223
Reception center, 16–18, 21
Rehnquist, William, 210, 212
Relford v. Commandant, 211
Reserve Officers' Training Corps (ROTC), 156–170; academic rigor, lack of, 161–162; founded, 156; personality of cadets, 163
Respondeat Superior, 160
Rickover, H. G., 166
Ridgway, Matthew B., 247–249
Roosevelt, Franklin D., 2, 120
Root, Elihu, 177
Rules of engagement (ROE), 250

Sadler, Barry, 14, 58
Saint Anselm's College, 10–11, 14, 26–27, 29, 42, 66, 154
Sarkesian, Sam, 244

Savage, Paul, 3
Schlesinger, Arthur Jr., 189
Schurz, Carl, 11–12, 241
Schwarzkopf, Norman, 243
Separation of powers in U.S. government, 216
Seven Days in May (novel), 216
Sheen, Fulton J., 87, 132–133
Silber, John, 244, 250
Skelton, Ike, 175–178, 193–194
Socrates, 9, 160
Solorio v. U.S., 211
Star-Spangled Banner, 252
Stockdale, James B., 168–169, 195, 232
Sullivan, Gordon R., 175
Surety testing, 104

Tactical officers, 40–51
Taylor, Maxwell, 97
Temporary soldier (professional temporary and career temporary), 86–94
Texas A & M University, 157
Thucydides, 225
Toner, Christopher H. (graduates from Airborne School), 63
Toner, Patrick J. (graduates from BCT), 34–35.
Traditionalist model, 233–238
Tribute to USAREUR/Vietnam Vets, 150–151
Tuchman, Barbara, 175
"Turning blue" in OCS, 55–56

Uniform code of military justice (UCMJ), 207–215
USAREUR (US Army, Europe), 68–150

van Creveld, Martin, 190–193
Vann, John Paul, 6
Vietnam veterans, 150–151
Vietnam war, 13, 203
Vinson, Fred M., 208, 213
Virginia Military Institute (VMI), 155–158, 177, 194
Virginia Polytechnic Institute and State University (VPI), 157

William and Mary, College of, 154
War crimes, 214–215
War Powers Bill, 216–217
Wells, H. G., 184
Westmoreland, William C., 220–221
Wiener, F. B., 214
Wife battering by platoon sergeant, 100–102
Wilson, Woodrow, 242

Xenophon, 37

About the Author

JAMES H. TONER is Visiting Professor of International Relations at the U.S. Air War College at Maxwell Air Force Base in Alabama and Associate Professor of Political Science at Norwich University in Vermont. He is a graduate of Infantry OCS and U.S. Army Airborne School. From 1968 to 1972, Toner was on active duty as a nuclear weapons officer in the U.S. Army, Europe. Recognized for his excellence in teaching, he has recently completed another book, *The Sword and the Cross* (forthcoming from Praeger in 1992), and has contributed to such journals as *Parameters, Military Review, Armed Forces & Society, Army,* and *Choice.*